PREPARING FOR EL NIÑO

HEARING

BEFORE THE

SUBCOMMITTEE ON ENERGY AND ENVIRONMENT

OF THE

COMMITTEE ON SCIENCE

U.S. HOUSE OF REPRESENTATIVES

ONE HUNDRED FIFTH CONGRESS

FIRST SESSION

SEPTEMBER 11, 1997

[No. 29]

Printed for the use of the Committee on Science

U.S. GOVERNMENT PRINTING OFFICE

45–763CC　　　　WASHINGTON : 1997

For sale by the U.S. Government Printing Office
Superintendent of Documents, Congressional Sales Office, Washington, DC 20402
ISBN 0-16-056044-6

COMMITTEE ON SCIENCE

F. JAMES SENSENBRENNER, JR., Wisconsin, *Chairman*

SHERWOOD L. BOEHLERT, New York
HARRIS W. FAWELL, Illinois
CONSTANCE A. MORELLA, Maryland
CURT WELDON, Pennsylvania
DANA ROHRABACHER, California
STEVEN SCHIFF, New Mexico
JOE BARTON, Texas
KEN CALVERT, California
ROSCOE G. BARTLETT, Maryland
VERNON J. EHLERS, Michigan**
DAVE WELDON, Florida
MATT SALMON, Arizona
THOMAS M. DAVIS, Virginia
GIL GUTKNECHT, Minnesota
MARK FOLEY, Florida
THOMAS W. EWING, Illinois
CHARLES W. "CHIP" PICKERING,
  Mississippi
CHRIS CANNON, Utah
KEVIN BRADY, Texas
MERRILL COOK, Utah
PHIL ENGLISH, Pennsylvania
GEORGE R. NETHERCUTT, JR.,
  Washington
TOM A. COBURN, Oklahoma
PETE SESSIONS, Texas

GEORGE E. BROWN, Jr., California RMM*
RALPH M. HALL, Texas
BART GORDON, Tennessee
JAMES A. TRAFICANT, Jr., Ohio
TIM ROEMER, Indiana
ROBERT E. "BUD" CRAMER, Jr., Alabama
JAMES A. BARCIA, Michigan
PAUL McHALE, Pennsylvania
EDDIE BERNICE JOHNSON, Texas
ALCEE L. HASTINGS, Florida
LYNN N. RIVERS, Michigan
ZOE LOFGREN, Califomia
LLOYD DOGGETT, Texas
MICHAEL F. DOYLE, Pennsylvania
SHEILA JACKSON LEE, Texas
BILL LUTHER, Minnesota
WALTER H. CAPPS, Califomia
DEBBIE STABENOW, Michigan
BOB ETHERIDGE, North Carolina
NICK LAMPSON, Texas
DARLENE HOOLEY, Oregon

TODD R. SCHULTZ, *Chief of Staff*
BARRY C. BERINGER, *Chief Counsel*
PATRICIA S. SCHWARTZ, *Chief Clerk / Administrator*
VIVIAN A. TESSIERI, *Legislative Clerk*
ROBERT E. PALMER, *Democratic Staff Director*

SUBCOMMITTEE ON ENERGY AND ENVIRONMENT

KEN CALVERT, California, *Chairman*

HARRIS W. FAWELL, Illinois
CURT WELDON, Pennsylvania
DANA ROHRABACHER, California
STEVEN H. SCHIFF, New Mexico
VERNON J. EHLERS, Michigan
MATT SALMON, Arizona
MARK ADAM FOLEY, Florida
PHIL ENGLISH, Pennsylvania
TOM A. COBURN, Oklahoma

TIM ROEMER, Indiana
PAUL McHALE, Pennsylvania
MICHAEL F. DOYLE, Pennsylvania
DARLENE HOOLEY, Oregon
RALPH M. HALL, Texas
EDDIE BERNICE JOHNSON, Texas
ZOE LOFGREN, California
ALCEE L. HASTINGS, Florida

---

*Ranking Minority Member
**Vice Chairman

CONTENTS

PREPARING FOR EL NIÑO

THURSDAY, SEPTEMBER 11, 1997

U.S. HOUSE OF REPRESENTATIVES,
COMMITTEE ON SCIENCE,
SUBCOMMITTEE ON ENERGY AND ENVIRONMENT,
Washington, DC.

The Subcommittee met, pursuant to notice, at 10 a.m. in room 2318, Rayburn House Office Building, Hon. Ken Calvert, Chairman of the Subcommittee, presiding.

Chairman CALVERT. The hearing will come to order.

This hearing of the Energy and Environment Subcommittee will come to order. Today we will look at the weather phenomenon known as El Niño. What was once thought of by Peruvian fishermen, who gave El Niño its name, as a localized freak weather pattern, we now know to be a change in the atmosphere over the Pacific Ocean that has a major effect on the world's weather.

In its simplest explanation, El Niño, which occurs every 2 to 7 years, is an abnormal warming of the eastern Pacific waters that interferes with the normal trade wind patterns. The ripple effect from El Niño has been held responsible for everything from droughts in Asia to floods in my home State of California.

As we will hear from testimony today, the phenomenon is not new, but the ability to predict its comings and goings is relatively recent. This year, for the first time, scientists are willing to be specific not only about the occurrence of El Niño but what effects on the weather it will have this coming winter in the United States and around the world.

If the predictions prove accurate, we will see a very wet winter in California, a pleasant mild winter on the east coast, abnormally mild and dry conditions in the Great Lakes States. Just how great an effect this El Niño event could have is reflected in the winter of 1982 and 1983 when California mud slides, crop failures in other parts of the world, and a general reversal of weather patterns was held responsible for numerous deaths and $8 billion worth of damage world-wide. Whether this year's event will be that serious is an open question.

This Subcommittee and the Committee on Science have given strong support to research associated with El Niño through funding of the interannual and seasonal climate research program, despite tight budget restrictions. H.R. 1278, the NOAA reauthorization bill—authorization bill—which passed the Science Committee on a unanimous vote, funds this program at almost $13 million in Fiscal Year 1998 and fully funds the requested $4.9 million for the tropical ocean and global atmosphere observing system known as

TOGA, which has also played a key role in gaining a better understanding of El Niño.

There has been some concern expressed by researchers over the funding levels for interannual and seasonal programs in the House Appropriations bill and how that affects TOGA. I can announce this morning that the Chairman of the House Appropriations Subcommittee, Mr. Rogers, and I have agreed to work in conference to assure continued funding for TOGA. And this will be outlined in a colloquy with myself and Mr. Rogers when the Commerce State Justice Appropriations bill hits the Floor, which I hope is today or early next week. If the current predictions come true, the investment in these programs will pay for themselves many times over.

At this hearing, we will look at both the state of the science of El Niño prediction and the reaction of the federal and state agencies that will have to deal with the consequences of extreme weather conditions. Among the issues to explore are: the state of confidence of predictions for weather conditions this winter; what state, federal, and local agencies are doing to prepare, and how much coordination is among these agencies; and is there as much danger of overreacting as underacting to these predictions.

Before I introduce our first panel, let me turn to my friend from Indiana, the distinguished Ranking Minority Member, Mr. Roemer, for his opening remarks.

Mr. ROEMER. Thank you, Mr. Chairman. I appreciate you holding this hearing today. And I appreciate our panel of experts to talk about El Niño.

I think it was Mark Twain who once said that everybody talks about the weather, but nobody can ever do anything about the weather. Our title of this hearing today suggests otherwise, that we can prepare for El Niño, that we can study El Niño, that we can better predict, prepare, study, and then get ready for the possible impact of a weather system that created havoc in the United States, particularly in California, back in 1982 and 1983.

We are not alone in our attempt to prepare and come to grips with the El Niño phenomenon. In the Indonesian coffee-growing region of southern Sumatra, tens of thousands of people have held prayer meetings for an end to an El Niño-induced drought threatening the world's biggest coffee crop.

In the northern Philippines, farmers have been carrying images of the Christ child into the rice paddies and corn fields to pray for rain after the drought that began in May.

In Australia, farmers are letting livestock loose in parched, withering wheat fields. The official estimate for the Australian wheat crop has been slashed by 28 percent.

In Africa, weather experts opened a 5-day conference in Zimbabwe this past Monday to discuss the El Niño weather pattern that has wreaked havoc across the globe and is expected to bring drought to that region.

Commodity brokers around the world are gambling enormous fortunes based on their best guesses regarding the El Niño effects.

What are we in the United States doing? As we will hear today, we have carried out a systematic and very productive research program over the past decade to help us understand and monitor El

Niño climate changes. We have with us today some outstanding scientists who have participated in that research.

What is needed at this point is to capitalize on that investment and transform this research capability into an operational predictive capability. Users in the transportation, agriculture, utility, and urban planning sectors do not want to read scientific journals. They want a clear, reliable, and timely forecast of the climate just as they receive forecasts of the day-to-day weather.

We are on the verge of establishing that capability. Unfortunately, the House Appropriations Subcommittee that has jurisdiction over NOAA does not apparently share this concern. The Commerce State Justice bill that will come to the Floor probably next week has eliminated the very funding that was proposed to establish this capability.

What is the value of an El Niño forecast? With world-wide economic losses in excess of $8 billion for the 1982–1983 event, the benefits of an efficient reliable forecast will be enormous. In the United States alone, the benefits versus cost for the forecasting capability proposed by NOAA is about 300 to 1 using the figures our witnesses will provide us with today.

We have always had bipartisan cooperation with NOAA. We have bipartisan cooperation with our modernization efforts. I am hopeful that we will have bipartisan cooperation with this predictive capability with El Niño.

Finally, I would suggest to our citizens and our legislators on Capitol Hill, that the cost of our emergency supplementals each year in this body that we must consider on the House Floor are enormous. Each year we appropriate hundreds of millions of dollars to different regions of our country based on weather climate that has delivered devastating results. Rather than continuing to fork out these billions and billions of dollars year by year, I would hope that we would improve our predictive capabilities to help our farmers, to help our agriculture, to help our businesses, to help our people, to help our children getting ready to be on school busses in the morning be prepared for what appears to be yet enormous consequences coming from another El Niño.

So I am very interested in what our panel here has to say this morning, very interested in seeing how the Congress will react to this next week when we take up this important legislation. And I know that the Congresswoman from California, Zoe Lofgren, will be hopefully joining us today and is interested in possibly offering an amendment on the Floor next week to address this situation as well.

So with that, Mr. Chairman, I look forward to this hearing and again want to thank you for your interest in this and you calling the hearing this morning.

Chairman CALVERT. In this instance and many other instances, my good friend Mr. Roemer and I will be working together to make sure that money is in the Appropriations bill. As a matter of fact, I will have a colloquy with Mr. Rogers. It has already been arranged and he has agreed to put that funding in. So, more than likely, an amendment will not be necessary.

Mr. ROEMER. Mr. Chairman, has he agreed to restore the full funding rather than——

Chairman CALVERT. He has agreed to work with us to make sure—to assist us in making sure the money would be restored.

Mr. ROEMER. The full funding that was cut?

Chairman CALVERT. What we are looking for is the full funding.

Mr. ROEMER. Okay. Well, I look forward to working with you and talking with you more about this.

Chairman CALVERT. Sure.

Panel I

Chairman CALVERT. Our first panel will look at the state of the science of El Niño prediction. Dr. J. Michael Hall is Director of Global Programs for the National Oceanic and Atmospheric Administration; Dr. Tim Barnett is a Research Marine Physicist with the Scripps Institution of Oceanography in La Jolla, California; and Dr. Andrew Solow is Director of Marine Policy Center at Woods Hole Oceanographic Institute in Massachusetts.

Would you please all rise and be sworn in? You will just rise and raise your right hand.

Do you swear to tell the truth, the whole truth, and nothing but the truth, so help you, God?

[All witnesses respond affirmatively.]

Chairman CALVERT. Gentlemen, without objection, your full written testimony will be included in the record. I would ask that you summarize your statements and remarks in 5 minutes so we will have plenty of time for questions.

So with that, Dr. Hall, your opening statement?

TESTIMONY OF J. MICHAEL HALL, DIRECTOR, OFFICE OF GLOBAL PROGRAMS, NATIONAL OCEANIC AND ATMOSPHERIC ADMINISTRATION, WASHINGTON, DC

Mr. HALL. Thank you, Mr. Chairman.

It is a pleasure to be here this morning to address a topic that is very important to my agency.

The hearing you are holding today I think would be important under any circumstances, and it is doubly important that it is happening while a major event unfolds, an event which we might anticipate in ways we have never been able to do before.

I have introduced into the record a lengthy statement of the state-of-the-art in El Niño forecasting. I will try in just a few minutes to summarize the essence of my testimony and with your permission call your attention to one or two important points in the summary of my testimony.

It is clear, I think, to most citizens of America that variable patterns of weather affect our daily lives. Some 15 years ago, scientists began to believe that the El Niño phenomenon and its atmospheric counterpart—this Southern Oscillation—was perhaps the most important mechanism affecting our lives, affecting weather patterns on time scales of a season out to a few years. Scientists in this and several other countries began a massive effort, consistently supported for 15 years by this Committee.

The result of that effort has led to some remarkable improvements in our understanding of the climate system and our ability to forecast El Niño events. We can, for example, now project the basic stage of El Niño in the tropical Pacific out to a season or several seasons in advance with useful skill. Secondly, we can calculate, at least in broad outline, the global atmospheric implications of this tropical phenomenon which we project out a season or two.

During this period of research—particularly the last several years—we have begun to learn how to use this information in selected decision frameworks around the world, working with other scientists in other countries as well as decisionmakers in the United States. That process of learning how to use the information has just begun. But nonetheless, important demonstrations of the relevance of today's predicted information I believe have already been made. A few of them are outlined in my testimony.

If I could focus on two important caveats in presenting to you our successes in predicting El Niño. The first of them is that in any variable pattern of weather over North America, El Niño is only part of the story. So however well we might be able to predict El Niño over North America, important cautions have to be introduced that there is a great deal of work to be done on other phenomena that affect patterns of weather over North America and that we are poised to conduct that research.

Secondly, I want to emphasize in particular that how to use El Niño forecast information in decision frameworks is itself the subject of fairly intensive research. It is not entirely obvious in any given decision framework—from agriculture to transportation to energy to foreign affairs—exactly how forecast information can be best made use of, so that a fairly intensive effort, working with our partner agencies, is needed to illuminate the value of the information. Again, we are poised to do that.

As you suggested, Mr. Chairman, in your introduction then, the NOAA effort is poised at a crossroads that has two divergent pathways. One is to exploit what we presently know how to do, to take our present research capability and make it operational in the agency. We have begun to do that. The appropriations issue that you mentioned—and of course your remarks about the appropriation process this year are most welcome—that issue has fundamentally to do with our capacity to take what we know how to do in El Niño forecasting into an operational mode and make these results routinely available to the American public.

At the same time, the second path we must take and are prepared to take is to learn how to do the remainder of the problem that we don't presently know how to do. It is becoming clear that El Niño—this fundamentally important phenomenon with its global manifestation—is one of a family of types of variability in the system. We are starting to get acquainted with the other members of that family. We don't know them nearly as well as we know El Niño, but we hope in the next 5 to 10 to 15 years to be back with you with stories as effective about these other types of variability as our present story about El Niño.

NOAA, then, has to pursue both of these paths, exploit what we now know how to do, and finish solving the variability problem. The President's budget that was submitted to the Congress reflects a dual character in NOAA's program on each of these paths.

I am very happy to hear that supporting both of those is a priority to this Committee. The observational effort that was mounted over the last 15 years in pursuit of the research I have described is fundamental to both of those paths. It is fundamental to understanding these other modes of variability, to improving El Niño forecasts. But it is also fundamental to the operational prediction effort which NOAA has begun and hopes to sustain.

Let me summarize, Mr. Chairman, by suggesting that the last 15 years of research have been scientifically rewarding. They have been already societally relevant and can be made more so. And they are of fundamental interest to NOAA in pursuing its program of services to the Nation. I suspect—truly believe—that the next 10 to 15 years will be every bit as rewarding as we realize fully our research investment.

I would be happy to take any questions you might have, Mr. Chairman. Thank you.

[The prepared statement of Mr. Hall follows:]

Testimony of
J. Michael Hall
Director, Office of Global Programs
National Oceanic and Atmospheric Administration
U.S. Department of Commerce

Before the
Subcommittee on Energy and Environment
Committee on Science
U.S. House of Representatives
September 11, 1997

Mr. Chairman and Members of the Subcommittee:

Fifteen years ago a small group of scientists and research managers based the development of a major research effort in climate on the premise that patterns of extreme weather were not totally random events and that the key to predictive insight lay in the interaction between the ocean and the atmosphere. That international multidisciplinary research effort today is the foundation underlying the National Oceanic and Atmospheric Administration's (NOAA) world-recognized advances in forecasting climate based on dynamical predictions of the stages of El Niño and the Southern Oscillation (ENSO) in the tropical Pacific. As we turn our attention toward improved understanding of the connections between ENSO events and longer term global climate changes, and how to incorporate predictive information into decision-making in climate-sensitive sectors, we stand to realize substantial economic gains associated with mitigating the costs of disaster and capitalizing on the advantages of a favorable climate where and when it can be anticipated.

The research program to which I refer, called TOGA for Tropical Oceans and Global Atmosphere, formally ended in 1994. By the time this program had matured it involved hundreds of investigators from more than eighteen nations. TOGA left a remarkable legacy, which includes:

1) a demonstrated ability to project basic El Niño conditions in the tropics a few seasons in advance with measurable skill;

2) a more limited but still useful ability to compute, at least in broad outline, the global atmospheric response which can be expected from projected conditions in the tropics;

1

3) an observational system, supported by several countries, that fully integrates space-based and in situ observations for monitoring El Niño conditions in the tropics and for initializing computer models;

4) a growing body of experience around the world in using forecast information in decision-making (and an increased appreciation of the subtle ways in which forecasts must be used and evaluated); and

5) a broadened program of research into El Niño and related processes to carry on in the aftermath of TOGA.

TOGA research in the United States was sustained by cooperative efforts in four agencies: NOAA, the National Science Foundation, NASA, and the Department of Defense's Office of Naval Research. It was formally incorporated into the U.S. Global Change Research Program in the late 1980s, where its legacies may be found today.

My testimony today is intended to express NOAA's sense of accomplishment at having led a highly successful research effort and, particularly, at having moved the research results into our operational activities in an efficient manner. But I must also emphasize that the job is not done. When we use today's models to peer into the future, what we see is extremely useful, but it is only a glimpse of the next season's climate compared to what we might do when our research investment is fully realized.

In brief, as we move aggressively to exploit what we have already learned for practical human benefit, we are also challenged to improve our basic ENSO forecasts, extend them to mid-latitudes more effectively and with higher resolution and, perhaps most importantly, place their predictive information in the context of the several other climate processes which, together, make up the variable patterns of weather we experience in our daily lives.

BACKGROUND

What is an El Niño and how is it related to the Southern Oscillation to form what we call ENSO? The climate system variability we call ENSO is a slowly changing oscillation (of irregular period) in the ocean circulation patterns of the tropical Pacific and the trade wind system which drives them. The atmospheric component of ENSO, the Southern Oscillation, is a slow variation in the sea-level atmospheric pressure difference between Darwin in Northern Australia and Tahiti in the central Pacific. This pressure difference is important because it forces the trade winds in the Pacific. Its variability corresponds to a variation in the winds blowing over the tropical ocean. The oceanic component of ENSO, El Niño, is a warming of the sea surface waters of the central and eastern tropical Pacific produced by gradual changes in ocean current patterns. The ocean currents that produce El Niño vary in response to the changing winds of the Southern Oscillation. Conversely, the redistribution of heat in the upper ocean by tropical ocean currents affects the trade wind system by changing the east-west atmospheric pressure forces which drive it.

In simple terms, the El Niño drives the trade wind system (measured by the Southern Oscillation) and the trade winds drive the ocean currents which produce El Niño (measured by the sea surface temperature in the eastern Pacific). The principal consequence of this "coupling" is a variation in rainfall patterns in the tropics. Huge, persistent patterns of rainfall normally seen over Indonesia move out into the central or eastern Pacific as the water warms there during an El Niño. When the El Niño fades, these rainfall patterns return to their normal position. The sea surface temperature in the eastern Pacific returns to normal (slightly colder than the surrounding waters) and the tropics is said to be in its "cold phase". Some call a well-developed cold phase "La Niña". The climate over much of the world is affected by ENSO simply because the varying position and intensity of tropical rainfall patterns change the prevailing winds over much of the planet.

Societal Costs Associated with Varying ENSO Conditions

Early-warning forecasts combined with insight from historical data afford opportunities for reducing vulnerability to ENSO-related climate fluctuations. The 1982-83 El Niño, which to date was the most intense El Niño in recorded history, and the 1992-95 period of repeated or prolonged El Niño conditions help illustrate some of the costs of interannual climate fluctuations which could be reduced or avoided with improved understanding and better forecasting:

• *Severe Storms along the West Coast* -- During the 1982-83 event, severe storms along the West Coast and associated high winds, coastal erosion, and flooding resulted in damages and losses of several hundred million dollars. Los Angeles had its wettest March since 1884. The storms caused major damage to fruit and vegetable crops, increasing prices substantially.

• *Flooding* -- Rainfall in Illinois during the first three days of December 1982 was nearly three times the normal amount for the entire month. Flood damage was estimated at $100 million. The inexorable lake-level rise of Great Salt Lake from 1982 to 1986 was at least partly related to the record rainfall and snowfall during the 1982-1983 El Niño. The rise in Great Salt Lake in the early to mid 1980s resulted in costly damage to many sectors of society, along with the disruption of local and interstate transportation. Interstate 80 was intermittently flooded and the roadbed was raised. The Union Pacific Railroad elevated its routes across the lake three times to keep the tracks above water. The international airport escaped inundation by 2.4 m (8 ft).

• *Drought in Midwest* -- A dry, hot summer followed the wet winter of 1982. The northern plains states experienced their second warmest August in more than 50 years. Corn crop yields were down 50 percent from 1982 levels. Crop losses were estimated at $10 billion. Fluctuations of this type remain a major research task because they are among the most difficult to forecast.

• *Increased Typhoons in Hawaii* -- During the 1982-83 El Niño, Iwa became the first typhoon to strike the Hawaiian Islands in 23 years with damages estimated at $234 million. Pacific storm paths are known to be directly affected by El Niño.

• *Disruption of Fisheries in West*-- In 1982-83 commercial salmon fishermen in Washington reported catches down more than 50 percent from expected numbers costing the fishing industry an estimated $400,000. Research indicates that this situation may be a case where El Niño enhances one or more longer term climate processes (mentioned later).

• *Mississippi Floods* -- The 1993 flood in the U.S. Midwest demonstrates vividly that social, agricultural and municipal planning which does not incorporate a recognition of the natural variability of climatic conditions can encounter costs sufficient to affect the region's economic base, perhaps even influencing national economic indicators. The devastation of the current flooding in the Midwest, which has been estimated at over $10 billion, is the result of natural variations within the climate system. The meteorological circumstances which created these conditions are poorly known, but NOAA's model experimentation and analysis of historical data suggest tropical influences at least in part. The historical record reveals that the strong El Niño events of 1972 and 1982/83 were also both associated with flooding of the upper and lower Mississippi respectively, causing billions of dollars in damages.

While some regions of North America experience negative impacts from El Niño, some areas benefit. For example:

• *Decreased Hurricanes on East Coast* -- During the recurring El Niño of 1992-1995, a persistent pattern of enhanced vertical wind shear and above-normal surface air pressure over the subtropical Atlantic inhibited tropical storm formation. With the evolution of ENSO into a cold phase in 1995, however, these factors both decreased, creating an environment conducive to the development of tropical storms and hurricanes. 1995 witnessed 19 Atlantic tropical storms, 11 of which attained hurricane status, making this the second (next to 1933) most active hurricane season since records began in 1871. The end of an unprecedented series of warming episodes in the Pacific set the stage for these storms.

• *Energy Savings for Northeast* -- In 1982-83 the northeast experienced a warm winter, reducing energy demands considerably. It has been estimated that the mild winter saved consumers approximately $2.5 billion in reduced heating costs.

• *Water Resources Management in Southwest* -- The southwest typically experiences enhanced streamflow and seasonal snow water runoff during the mature winter phase of El Niño, and it experiences diminished streamflow and snow melt during La Niña phases.

• *Increased Fish Catch in British Columbia and Alaska*-- In 1982-83, Canadian and Alaska salmon fisherman had bountiful catches, in contrast to the reduced catches in the Pacific Northwest.

Research Achievements 1982-1997

In a recent report of the National Research Council, *Learning to Predict Climate Variations Associated with El Niño and the Southern Oscillation: Accomplishments and Legacies of the TOGA Program* (1995), it was noted that, "As the TOGA decade closed, the improvement in our understanding of short-term climatic fluctuations on our planet are clearly evident....TOGA accomplished much in its decade, especially in observing, understanding and predicting ENSO in the tropical Pacific....TOGA went far towards fulfilling its goals -- perhaps further than anyone could have foreseen at the beginning of the program. There were many successes."

Since 1982, the National Oceanic and Atmospheric Administration has led the U.S. interagency program of research into interannual climate variability. Agency efforts in the United States were fully integrated into the World Climate Research Programme in order to bring the maximum resource to bear in achieving the program's objectives. During this period, NOAA, primarily though its Office of Global Programs, provided approximately 50 percent of the funding for the U.S. portion of this multi-disciplinary international program. Ultimately, more than 18 countries participated in the TOGA program; a comparable number now work together on follow-on research (discussed later). NOAA's major contributions to the research effort were:

• the development and oversight of a peer-reviewed grants program involving both NOAA and university scientists, which provided much of the intellectual horsepower for the advances in our understanding of ENSO processes;

• the development of an integrated interagency management framework, incorporating NRC scientific advice and oversight, which assured strong intellectual leadership and effective use of resources;

- the establishment of an observational system spanning 10,000 miles of the equatorial Pacific Ocean to support real-time monitoring and experimental prediction of ENSO;

- the development of a suite of coupled ocean-atmosphere models capable of predicting ENSO up to a year in advance; and

- a solid start on research directed toward the incorporation of ENSO forecast information into decision-making frameworks within a variety of different sectors of society and at locations throughout the world.

Much of TOGA's success can be attributed, in part, to the program's focus on a portion of the climate puzzle in which the pieces seemed likely to fit together. A great deal of research is still required to develop the skill for predicting short-term climate variations caused by other processes or in other parts of the world. The National Research Council report specifically recommends efforts to maintain the observing system on a multinational basis, to create a multinational institute for developing ENSO forecast models and stimulating applications of short-term climate forecasts throughout the world, and to sustain continued research on seasonal-to-interannual climate variations as well as the influences of longer timescales and their predictability.

NOAA's continuing emphasis on the research problem will ensure that the decade-long investment in TOGA is carried forward to ensure that seasonal to interannual forecasts are used by decision-makers for substantial economic and societal benefit.

Since the completion of TOGA, NOAA has capitalized on its successes by:

- initiating systematic climate forecasts on seasonal-to-interannual timescales that are being used by decision-makers in various sectors including agriculture, health, energy, coastal fisheries, forestry, transportation, and water resources;

- convening an international forum of 40 countries to launch the International Research Institute (IRI) for Climate Prediction called for in the National Research Council's report as an outgrowth of TOGA. The meeting was hosted by the President's Science Advisor, Dr. Jack Gibbons, and the NOAA Administrator, Dr. D. James Baker;

- accelerating research into the use of climate forecasts for societal and economic planning worldwide; and

- formulating a program of continuing research to address the challenges outlined in the NRC report.

Today's Advances in Climate Forecasting

Fifteen years ago, operational agencies like NOAA were incapable of issuing ENSO forecasts of pending warm or cold events. Due entirely to the international investment in research and observing systems described here, we are now able to predict such events three months to a year in advance and issue experimental regional climate forecasts. From the earliest days of NOAA's seasonal to interannual research program, a key goal has been the transfer of research knowledge and observing systems to operational climate services. While a major research effort is still underway and there is much room for improvement in forecasts, we have begun that technology transfer. In fact, in the FY 98 budget submission, NOAA proposes to transfer the ENSO Observing System from the research program to an operational status within the agency. Completion of this transition will represent a remarkable advance in NOAA's capacity to provide predictions of variability in weather patterns affecting US citizens.

The best demonstration of this new capability can be seen in the El Niño now underway. It was forecasted well in advance by NOAA's National Center for Environmental Prediction (NCEP). The most recent forecasts, issued in August, which can be found on the internet at **http://www.ogp.noaa.gov/enso,** suggest the following conditions:

El Niño Forecast: Since March 1997, strong warm episode (El Niño) conditions have developed in the tropical Pacific (See Figure 1). Sea surface temperatures throughout the equatorial east-central Pacific increased during April and May. By July, ocean surface temperatures were near record levels for the month in many sections of the equatorial Pacific. Departures from normal exceeded +3°C (+5°F) along the equator east of 140°W. and were greater than +5°C (+9°F) near the coasts of Ecuador and northern Peru. Over the past few seasons the NCEP and other model predictions have consistently indicated the development of a strong warm episode (See Figure 2). The August forecasts indicate that strong warm episode oceanic conditions, comparable to those observed during the 1982-83 El Niño, the most intense event of this century, will continue throughout the remainder of 1997 and into early 1998.

Regional Climate Forecasts: Based on the NCEP sea surface temperature (SST) predictions and information generated by historical studies on the effects of warm episodes, we expect during the northern winter season wetter than normal conditions to prevail over most of the extreme southern United States. Temperatures are likely to be warmer than normal in the northern half of the United States and along the California coast, and slightly cooler than normal along the Gulf Coast.

Regional forecasts also project conditions for the rest of the world. We expect drier than normal conditions to persist over Indonesia and eastern Australia during the next several months. Drier than normal conditions are also likely over most of Central

FIGURE 1

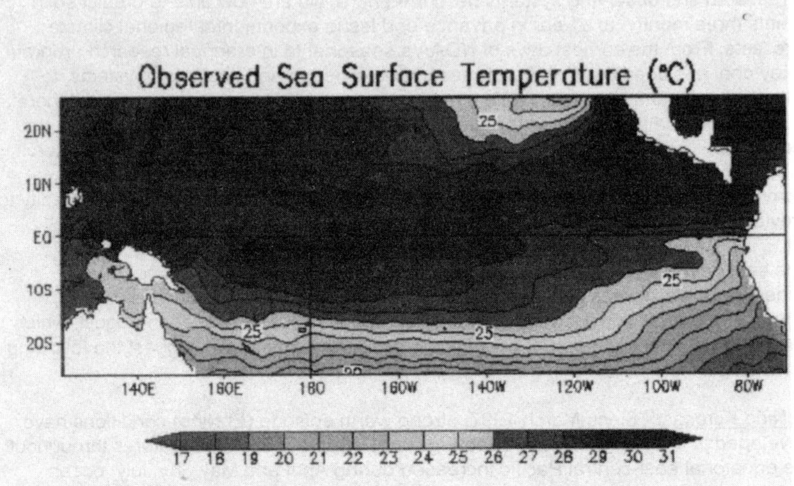

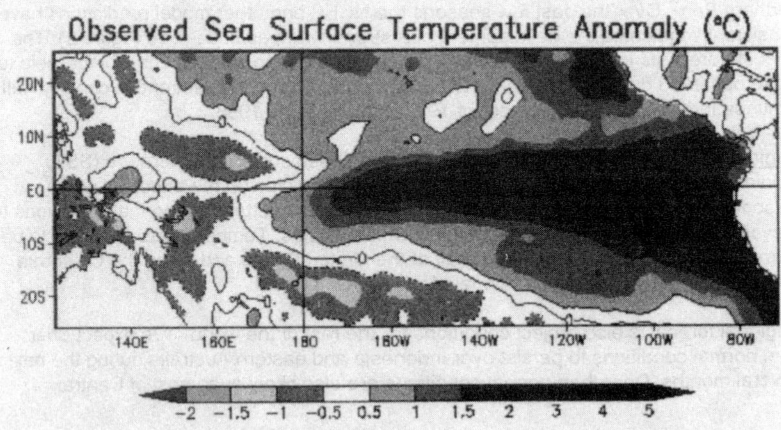

7-day Average Centered on 03 September 1997

FIGURE 2

MULTIVARIATE ENSO INDEX for the 6
strongest historic El Niño events vs. current event
Standardized Departure
57+
65+
72+
82+
86+
91+
97+
3
2
1
0
-1
-2
86/87
91/92
57/58
82/83
65/66
72/73
Dec/Jan
Mar/Apr
Jun/Jul
Sep/Oct
Dec/Jan
Mar/Apr
Jun/Jul
Sep/Oct
Dec/Jan

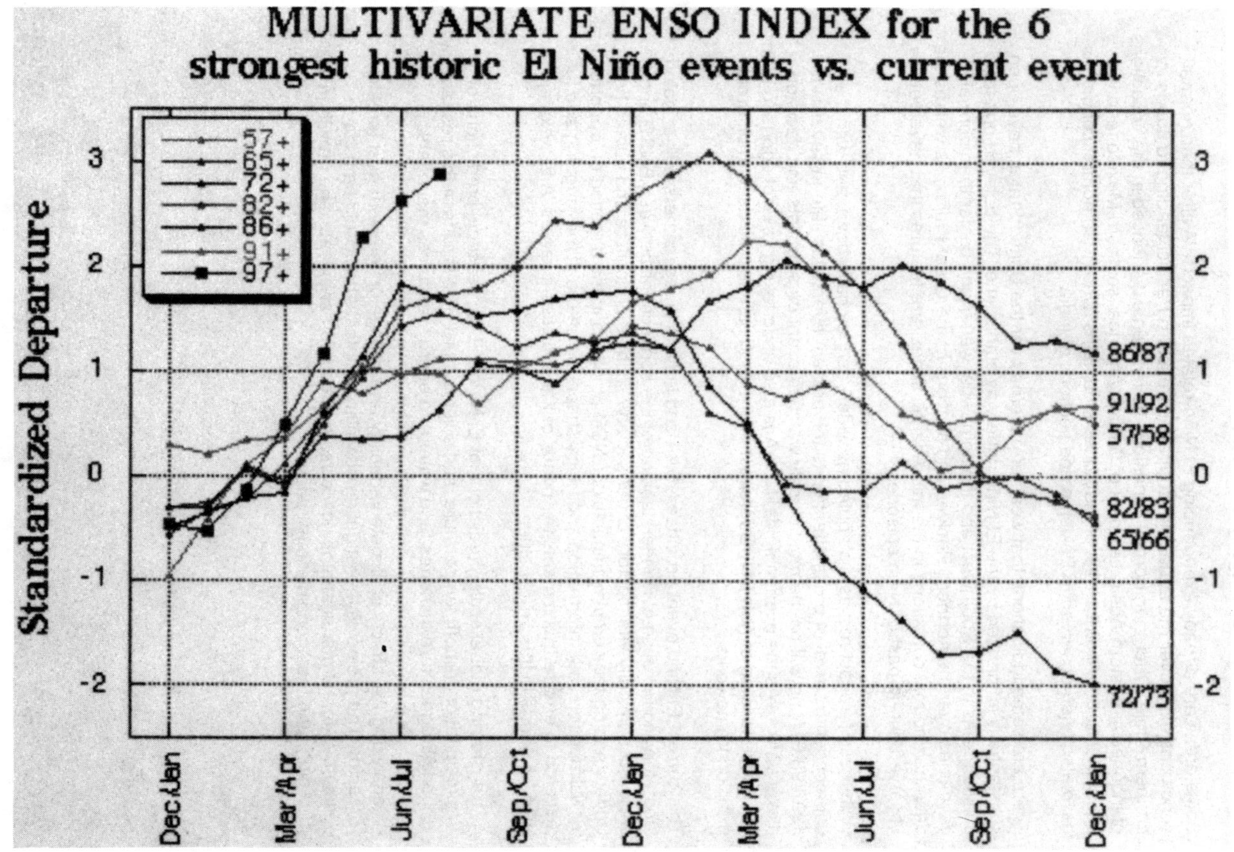

America and the Caribbean Sea. Rainfall should continue to be heavier than normal over the central and eastern equatorial Pacific and from central Chile eastward across northeastern Argentina, Uruguay and southern Brazil.

Some areas are already experiencing El Niño-related impacts. Wetter than normal conditions exist over the equatorial central and eastern Pacific, leading to floods in northern Chile. Drier than normal conditions have been experienced in many sections of Indonesia, eastern Australia, most of Central America, southern Mexico, and in equatorial South America east of the Andes Mountains. Drier than normal conditions have also been observed in Pakistan and sections of India.

There are some indications that weather patterns over the United States have been substantially influenced by the El Niño underway at this time. The delayed onset of summer rains in the Southwest and the somewhat wetter and cooler than normal conditions over the northern Rockies and sections of the Great Plains, as well as the drier than normal conditions in the mid-Atlantic states, are features that have been observed during past El Niño episodes.

With these forecast results distributed as widely as they have been, and with the unprecedented level of press attention to the present event, it is safe to say that the 1997-98 El Niño will be the most closely watched event by scientists and decision-makers alike. As the event unfolds, NOAA is also undertaking several efforts to further our understanding of how ENSO forecast information can best be used in different decision frameworks.

This current El Niño event provides NOAA and university researchers with a "natural laboratory" to increase our understanding of the connections between ENSO and socio-economically vital sectors of our society, such as agriculture and human health. Understanding the complex impacts of ENSO points us in the direction of opportunities for utilizing forecast information and helps shape the nature of the ongoing and future research to ensure that it meets the needs of the societies it serves. A few examples of NOAA's efforts in this area include:

- transferring global scale predictions of El Nino to U.S. seasonal predictions of precipitation, runoff, temperature, and soil moisture. The results of this research will benefit agriculture, fisheries, and water and energy resources managers;

- analyzing the effects of climate variability and change on the Northeast forestry sector. The project is investigating how climate information could be used to improve forest management; how climate affects forest dieback and growth in the region; and the economic implications;

- analyzing the effects of El Nino and La Nina events on crop yields across the U.S. in order to provide information useful to the farming sector;

- conducting a small pilot project whereby stakeholders throughout California will be provided with regional El Nino forecast information this Fall and will document their responses to the forecast with particular attention to changing operational practices that have financial implications. This exercise will provide insight into the process whereby decision-makers become familiar with new types and sources of probabilistic information in the context of the range of factors that affect their activities;

- advising the Agency for International Development regarding probable climate patterns around the world. Early warning climate information is essential to famine early warning and foreign disaster assistance;

- organizing selected researchers around the world in public health, climate and ecology to study the human health impacts of climate variability using this El Niño as a "natural laboratory". This pilot activity will have a particular focus on mosquito-borne, rodent-borne, and water-related infectious diseases. One of the geographic areas to be studied is the Atlantic seaboard from Labrador to the Caribbean Sea;

- working with scientists and public officials from several regions of South America to conduct a series of activities to assess the impacts of the current El Nino and to explore opportunities to incorporate this information into decision making frameworks; and

- organizing a forum entitled the *Southern African Climate Outlook* in Zimbabwe in September 1997. The meeting will focus on building a consensus outlook for the El Niño during the upcoming rainy season (November-April). A mid-season correction in December will adjust the predictions made the previous September and produce a revised outlook for the remainder of the rainy season. A post-season assessment is scheduled for April 1998 to evaluate the success of previous Outlooks.

The Importance of the ENSO Observing System

The in situ observing system in the tropical Pacific (See Figure 3), which provides essential upper ocean and surface measurements for model-based forecasts of the ENSO phenomenon, is a critical underpinning of any climate forecast system in the United States. This observing network is run by the NOAA Environmental Research Laboratories in collaboration with university partners and in cooperation with other countries. NOAA's laboratories designed and developed state-of-the art, low-cost, efficient instruments to gather important data on surface winds, atmospheric pressure,

FIGURE 3

The ENSO Observing System

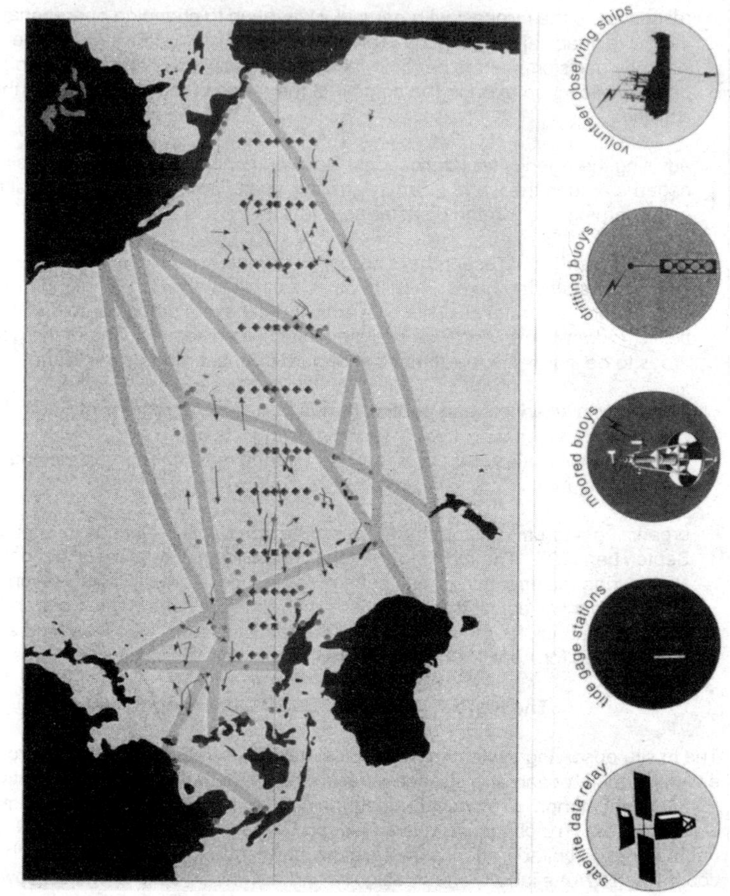

temperature, and ocean currents. Data from the network are available in real time on the World Wide Web at www/pmel.noaa.gov/toga-tao/el-niño/ and are used extensively by researchers as well as operational forecasters worldwide. This state-of-the-art technology is designed to be transferred to other locations where surface observations are critical to improving our understanding of climate processes and to enable regional climate forecasts. Several efforts to transfer the technology to other nations are presently underway.

The ENSO Observing System components were developed primarily through research programs designed to better understand the coupled ocean-atmosphere system in the Pacific. NOAA is now committed firmly to transforming these proven research observations into an operational system to provide routine observations for an end-to-end operational climate forecast system to serve a broad spectrum of U.S. decision-makers.

The ultimate benefits of the observing system are those that accrue to society based on responding to, or mitigating, the impacts of seasonal climate variability, such as floods, droughts, and abnormally cold or warm seasons, in sectors including energy, water resources, transportation, and agriculture. The observational system is also critical in providing data sets to be used by both federal and academic researchers to refine our understanding of the climate system and to improve ENSO forecasts. NOAA believes that the projected benefits for the public far outweigh the cost.

Research: The Next Decade

The TOGA program accomplished much in its decade, especially observing, understanding and predicting ENSO in the tropical Pacific. Many questions about ENSO and other types of interannual variability around the world, especially in the middle latitudes, remain exclusively, or in part, unanswered. Future research (See Figure 4) should focus on the following:

- continued long-term observations and improved experimental ENSO predictions in the tropical Pacific Ocean;

- expanded long-term observations in other tropical regions to encompass the major precipitation regimes of the globe, including the monsoon regions of the Americas, Africa, and Asia-Australia;

Future Research

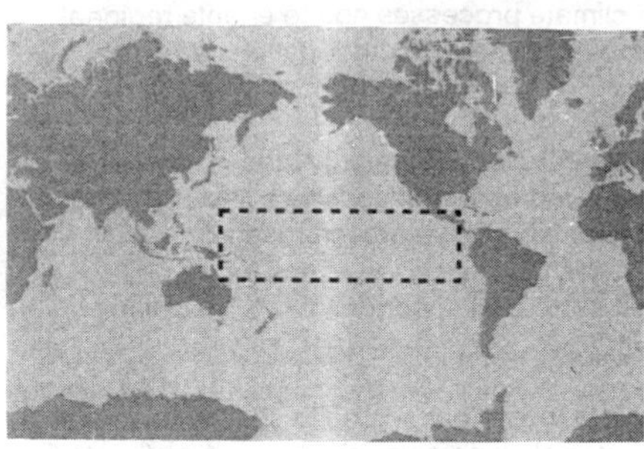

1. Continue long-term observations and experimental ENSO prediction in the tropical Pacific Ocean.

2. Expand long-term observations to the other tropical oceans to encompass the major precipitation regimes of the globe including the monsoon regions of the Americas, Africa and Asia-Australia.

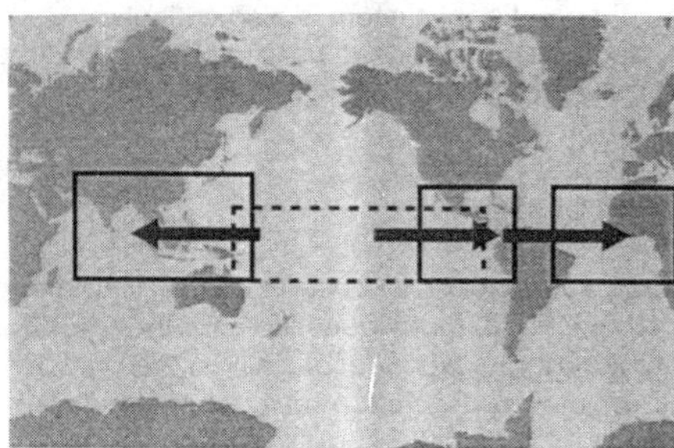

3. Improve predictability of global temperatures and precipitation on time scales of weeks to years and decipher how temperature and precipitation patterns interact and influence each other .

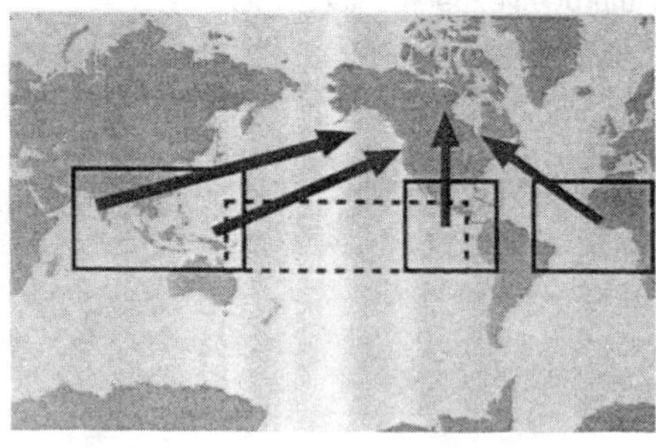

4. Determine the influence that the variations in the climate of the tropics have on the temperature and precipitation patterns of North America.

- improved predictability of global temperature and precipitation on timescales of weeks to years, based in part on knowing how temperature and precipitation patterns interact and influence each other; and

- new scientific insight and numerical models to project, with far greater precision, the influence that ENSO and other variations in the climate of the tropics have on the climate of higher latitudes, particularly over North America.

To address these issues the international climate research community has proposed to undertake several major research projects to improve our understanding of ENSO and the American and Austral-Asia monsoons, which are essential to our fundamental understanding of climate variability over North America. NOAA is playing a major role in addressing these problems.

While TOGA made possible our ability to forecast El Niño up to a year in advance with useful skill, the forecasts are limited in that they focus on the evolution of the tropical Pacific and its related climate impacts. Forecast skill is highest in the tropics near the source of El Niño and diminishes at higher latitudes (e.g. over North America) where other processes frequently play a greater role.

NOAA's Global Ocean Atmosphere Land Systems (GOALS) program was designed to continue the research needed for continuous improvements of El Niño predictions and to extend predictability of climate fluctuations beyond the tropical Pacific to include the effects of the other tropical oceans, higher latitude oceans, and land surface processes on seasonal-to-interannual climate variability, particularly at higher latitudes.

Another important component of the GOALS program is the Pan-American Climate Studies (PACS) Program. The overall objective of this international initiative is to extend the scope and improve the skill of operational seasonal-to-interannual climate prediction over the Americas. Particular emphasis is placed on summertime rainfall, for which reliable seasonal forecasts do not yet exist. In the context of PACS, climate prediction is focused not only on seasonal rainfall and temperature, but also on the frequency of extreme weather events such as hurricanes and floods over the course of a season or seasons.

Another important international research effort is conducted within the international Global Energy and Water Cycle Experiment (GEWEX) of the World Climate Research Programme. NOAA is focusing its GEWEX efforts on the Mississippi River Basin through the GEWEX Continental-scale International Project (GCIP). GCIP was developed to determine the variability of the Earth's hydrological cycle and energy exchange budget over a continental scale run-off basin; to develop and validate techniques for coupling atmospheric and surface hydrological processes in climate models; and, to provide a basis for translating the effects of future climate change to impacts on regional water resources.

NOAA's GEWEX and GOALS programs are combining their efforts to initiate a North American Monsoon System Study which will represent the only concentrated U.S. research effort to improve models for forecasting seasonal-to-interannual precipitation anomalies, like those in the Midwest in 1993, seasons to years in advance.

Underlying these research efforts is an international observational system that extends around the equator. The first proposed extensions of the existing Pacific system are into the subtropical eastern Pacific Ocean and the equatorial Atlantic Ocean. Eventually, and as our resources allow, observational programs will be mounted in the Indian Ocean and the full Pacific Ocean, with strong support from agencies and scientists in other nations.

Mr. Chairman, it might be constructive to examine briefly one or two of the larger lessons of ENSO research; I believe they have implications for this hearing. Perhaps the most fundamental lesson is that there is some inherent predictability in the climate system and we can hope to discover it and to exploit it through carefully targeted research. The climate system is not entirely random and fundamentally unpredictable in its behavior. It will have some degree of inherently random behavior, but some order will prevail in the system and can be understood. Our job as researchers is to remember the difference and to learn how to exploit the order we find in system behavior. El Niño's effect on the world's climate is partly predictable because the atmosphere acquires memory from the ocean in precisely that part of the world where the two can interrelate most effectively, in the tropics. What other processes effectively add memory to the system? We are pursuing GOALS because we believe that over much of the Earth's surface, the land surface itself adds memory to the atmosphere if we characterize it properly and model it well.

A second major lesson from ENSO suggests that the climate system has a few highly preferred "modes" of variability and that, over large portions of the world, some combination of these modes largely determines the predictable portion of the climate signal. ENSO would be one such mode; other tropical modes would include American monsoon processes, for example.

Very preliminary research results to date have suggested two additional modes of variability which may affect regional climate patterns over North America. These modes of variability seem to alter on larger timescales ocean and atmosphere behavior over large regions of the Northern Hemisphere in a way that changes weather patterns for twenty years or more. Together with ENSO they might provide substantial insight into how and why climate changes over North America from scales of season-to-season out to several decades. These modes, the Pacific Decadal Oscillation (PDO) which may affect patterns over much of the Pacific basin and nearby North America, and the North Atlantic Oscillation (NAO) which appears to influence climate over Europe and the Northeastern United States are very poorly understood at present. These might properly be the subject of a future research brief; for now it is sufficient to note that pursuit of a continuum of climate time scales, with research focused strongly on suspected "modes of variability," will be the essence of NOAA's research strategy for the coming decade.

The Value of Improved Climate Forecasts

Better understanding of the ENSO phenomenon will allow better resolution, longer lead time forecasts of variables such as precipitation, temperature, and snowpack in a region. Our improving ability to forecast these variables is beginning to provide decision-makers in climate sensitive sectors with a basis for adjusting behavior to minimize disruption or take advantage of beneficial climate. Early examples that attempt tp scope the value of these improved climate forecasts include:

• *Water Resources*--Reduced hydroelectric capacity in California can result in greater reliance on fossil fuels, thus affecting local pollution conditions. Every year water managers in California face the competing needs of low reservoir levels to guard against flooding versus storing enough water to meet the summertime water demands. The early-warning aspect of climate forecasts can contribute to the more efficient use of water and the anticipation of energy needs and allocations. A NOAA-funded (since FY1995) Columbia River watershed study is leading to a better understanding of the use of seasonal climate information in water management (hydroelectric power, flooding and coastal erosion, anadromous fisheries, multiple uses) and agriculture.

• *Fisheries*---A NOAA-funded study suggests that the ability to forecast the onset of an El Niño would help efforts to stem the decline of coho salmon on the West Coast. For example, in 1982 fisheries experts predicted that nearly 1.6 million wild coho salmon would return to spawn in Pacific Northwest streams in 1983. Only an estimated 667,000 showed up, or 42 percent of what had been expected. Incorporating ENSO forecasts in fisheries management could help to avoid extreme measures, such as closing the fishing season.

• *Agriculture:* Economic research shows that improved long-term forecasts of the ENSO phenomenon can result in economic benefits of $240-324 million per year to the agricultural sector in the U.S. The benefits occur when farmers make optimal planting, harvesting, and crop selection decisions based on predictions of the warm (El Niño), cold (La Niña), and normal phases of ENSO. For example, a NOAA-funded modelling study indicates that yields of important crops such as corn, wheat and soybeans can vary by as much as 15 percent to nearly 30 percent, depending upon the phase of ENSO.

We expect that further work in this area on the use of information will improve our ability to quantify the value of ENSO forecasts for the range of affected sectors. We are committed to supporting research that explores the nature of vulnerabilities to climate fluctuations as well as further experience in the actual incorporation of forecast information to decision-makers.

Mr. Chairman, that concludes my testimony. I want to thank you and the Committee for this opportunity to discuss ENSO research and forecasting. I cannot imagine a more timely and exciting scientific issue than the slowly unraveling mystery of climate system behavior and its effects on human societies. Our experience of the last 15 years in ENSO research has proven to be rewarding scientifically, of clear relevance societally, and of fundamental value to NOAA in its ongoing services to the Nation. I believe the future holds equal or greater promise for return on our research investment.

I would be pleased to answer any questions that you or other members may have.

BIOGRAPHICAL SKETCH

NAME: **J. Michael Hall**

POSITION: **Director, Office of Global Programs, NOAA**

BORN: May 28, 1943
                       Lusk, Wyoming

EDUCATION: B.A.,   Rice University, Physics - 1965
                       M.S.,   University of Washington, Physical Oceanography - 1968
                       Ph.D.,   University of Washington, Physical Oceanography - 1971

EXPERIENCE:

Mar.1986 - Present   Director, Office of Global Programs (previously Climatic and Atmospheric Research)
                       NOAA
                       Silver Spring, MD

Dec.1982 - Jan.1989   Director, U.S. TOGA (Tropical Oceans and Global Atmosphere) Projects Office
                       NOAA
                       Rockville, MD

Mar.1980 - Dec.1982   Program Manager, Ocean Climate Technology
                       Office of Ocean Technology & Engineering Services
                       NOAA
                       Rockville, MD

Oct.1977 - Mar.1980   Chief Scientist, Office of Ocean Engineering
                       NOAA
                       Rockville, MD

Feb.1970 - Oct.1977   Executive Scientist of the U.S. Committee for the Global Atmospheric Research Program
                       National Academy of Sciences
                       Washington, DC

Jan.1973 - Feb.1976   Chief Scientist, Data Buoy Office
                       NOAA
                       Bay Saint Louis, MS

Nov.1970 - Jan.1973   Research Scientist
                       Shell Development Company
                       Houston, TX

Jun.1968 - Nov.1970   Research Scientist
                       Boeing Scientific Research Laboratories
                       Seattle, WA

AWARDS: U.S. Department of Commerce Gold Medal (1989)
                       American Geophysical Union Ocean Sciences Award (1986)
                       American Meteorologist Association Special Award (1993)

OTHER ACTIVITIES: Vice-Chairman, Global Change Working Group of the Committee on Earth Sciences, 1987 - Present
                       Chairman, Board of Directors, NOAA Climate and Global Change Program, 1987 - Present
                       U.S. Representative to the Intergovernmental Board for TOGA, 1986 - Present
                       Co-chairman, NOAA Council for the Sub-Tropical Atlantic Climate Studies (STACS), 1980 - 1983
                       Chairman, Marine Board (National Academy of Engineering) Panel on Drifting Buoy Applications,
                           1979 - 1980
                       U.S. Representative for the International Southern Hemisphere Drifting Buoy Program of the Global
                           Weather Experiment (GARP), 1975 - 1977
                       Member, American Meteorological Society and American Association for the Advancement of Science

Chairman CALVERT. Thank you for your testimony, Doctor. Next, Dr. Tim Barnett. Doctor?

TESTIMONY OF TIM BARNETT, RESEARCH MARINE PHYSICIST, SCRIPPS INSTITUTION OF OCEANOGRAPHY, LA JOLLA, CA

Mr. BARNETT. Thank you. And it is also a pleasure and a privilege to be here.

I have two requests to make of you today—two suggestions—but first a little bit of background.

What do I do? Well, I do everything from detecting global warming signals to global sea level variations. But for the purposes of this meeting, I have spent a lot of my career trying to make El Niño prediction models and predicting what the impact of El Niños will be not only over the United States but globally.

Figure one in my testimony shows a forecast that we made in November that clearly shows an El Niño to be expected approximately a year later. Well, what is happening to that? Right now, temperatures in the equatorial Pacific, as you know, are near record heights. In other words, the forecast is turning out to be really quite well—maybe we were lucky. I don't know.

What is going to happen in the United States? You have already summarized that. My goodness, you know as well as I do. The one thing I would add in is that the northwest part of the United States is apt to be warm and dry—particularly dry—and you could see some problems in that region later.

I think the key question and the reason I am here—what are we going to do about this? What are we doing to prepare? I can tell you a few things that the Scripps Institution is doing, whom I am representing here today.

The sport fishing industry in California is a billion dollar a year operation. In January, I started working with them to let them know what was coming down the line at them. This is important for a whole bunch of investment reasons for them. I will tell you, I got a call earlier in the week. The sport fishing association could not be happier. They are having one of the best years ever. We can thank El Niño for almost all of that. I don't want to take the credit for that.

Scripps organized a forum. We invited people from cities and counties that we felt might be impacted. We expected 50 people; 300 showed up. They sold out our auditorium, there were 100 people standing out in front of it listening to the proceedings through a microphone.

As a result of that, we are working with Long Beach and Los Angeles now with some type of risk assessment programs. We are trying to come up with a mitigation strategy, helping them with the uncertainties in the forecast, which is a crucial thing because when I came back from the summer I found out that it was a given that California would be flooded away. I am sorry, that is really not the case here. The probabilities associated with that are maybe only about 70 percent.

On the other hand, the event that is underway now—we only have one analog of it in historical record and that was the 1982–1983 event. So I think prudence dictates that we prepare for the worst, although there is a probability that it won't happen.

One other thing we're doing at Scripps that is really fairly impressive—we have gotten together with our colleagues not only in Mexico, but up and down the entire west coast, and put together a response measurement strategy where we'll actually go out in the ocean in real time and make measurements of not only the physical properties but biological properties of the ocean—watch this whole El Niño develop. If the forecasts are right that we are going to switch over into a cold event next year, we have an unprecedented opportunity to watch this and try to understand how it evolves.

There is a problem with that program, and perhaps in Q&A you will give me a chance to tell you what it is because I think you and this Committee could help quite a bit there.

So that brings me to my first request. The Nation has spent a lot of years and a lot of money on people like me trying to develop a forecast technology. We need a national program to learn to apply that technology.

Mr. Roemer, I thought, made a very persuasive case for that in his opening statement. If we don't, it is foolish. It is like making an investment and never collecting the interest on your investment. So that is the first thing that I would like to request that you consider.

The second thing has to do with the TOGA TAO array—or the TOGA array—whatever you want to call it. I wasn't quite sure what you said in your statement there, if the funding is there or not. But we desperately need that array to continue in an operational mode for as far into the future as we can afford it.

I can tell you about how that is vital to forecasting. I can tell you how hundreds of scientists use the data from that array. We have learned things about the ocean we have never seen before. But to give you just a real concrete example in conclusion, I would like you to look at figure three of my written testimony. The title is the "Monthly Mean TAO Temperature Anomalies."

I indicated earlier that a forecast was made in November that an El Niño would occur. Let me show you how the TAO array has helped to verify that.

You see six panels there. Each one represents a snapshot of the tropical ocean along the equator from Asia to South America. At the top of each snapshot is the sea surface then down to 500 meters. What you can see in December is a little yellowish blob over in the western half of the Pacific Ocean. As you watch through January, March, February, you can watch the thing propagate across the ocean. Notice in the early part of the year there it doesn't appear at the surface. By March it is impacting the South American coast and another one has been generated, which eventually by the end of June down here you can see has a very strong warm signature.

Well, when I started talking to the fishing people in January, as I said, there was no indication that the surface there was going to be an El Niño. But an expert taking one look at these pictures says, "There it is. It is on the way." So the TAO array in this case would have given us an extra 3 or 4 months, adding confidence in the forecast. It would have given us a longer lead time to react had we chosen to do so.

I might add in the last panel you will see the blue is cold anomaly in the western Pacific again. That is a death sentence of the current El Niño that is etched into the thermal structure of the equatorial Pacific.

So again I think this speaks for itself. A TAO array is such a valuable tool for forecasting that I just can't emphasize it enough. All I can do is what happened. We can watch this unfold in real time, check out our computer model in real time, and give us confidence in the forecast.

Again, then, let me thank you very much for the privilege to be here today. I appreciate it. And like Mike, I would be glad to answer any questions you may have.

[The prepared statement of Mr. Barnett follows:]

Testimony of Dr. Tim Barnett
Scripps Institution of Oceanography
University of California, San Diego
Before the
House Science Subcommittee on Energy and Environment
September 11, 1997
"Preparing for El Niño"

Mr. Chairman and members of the Subcommittee, thank you for this opportunity to discuss the Scripps Institution of Oceanography's role in forecasting and preparing for the predicted El Nino. I am Tim Barnett, a researcher with U.C. San Diego's Scripps Institution of Oceanography Climate Research Division. On behalf of Scripps, I want to express my appreciation for your interest in El Nino prediction research.

Economic analyses and recent experience with prolonged weather-related extremes such as the Midwest drought of 1988, the California floods of 1994 and 1995, the active 1995 Atlantic hurricane season, and flood conditions across the country this year, demonstrate the significant social and economic benefits associated with the development and application of new capabilities to forecast climate conditions up to a year in advance. As I will describe later in my testimony, Scripps serves as host for the experimental forecasting division of the International Research Institute for Climate Prediction (IRI)--a newly established international scientific institution specifically created to provide experimental climate forecasts for use by affected communities around the world.

With support from NSF, NASA, NOAA, and DOE over that last 20 years, I have been working with colleagues from Scripps and other institutions to develop and improve our ability to make these forecasts. My work has been focused on seasonal-to-interannual prediction of El Niño and global climate and, more recently, prediction of decadal climate changes which may turn out to be more important than El Niño.

As you have heard from Mike Hall and in countless press articles and media reports over the past few months, scientists are predicting a significant El Niño (warm) event for 1997-1998. In fact, it is here now! My colleagues and I at Scripps first predicted this event in November of 1996 when our modeling results suggested a moderate warm event would occur during the winter of 1997-1998 (see Figure 1).

Ocean temperatures in the Eastern Equatorial Pacific (a primary indicator of an El Niño) have been warming as quickly as at any time in the historical record for this stage of an El Niño and the predictions indicate a little further warming. As you have heard today, scientists anticipate that the strength of the 1997-1998 El Niño will exceed the 1972-1973 event (the second largest of the century) and some scientists, including my colleagues at SIO and I, believe that this year's event may rival or exceed the 1982-1983 event--the strongest in recorded history.

Estimates of global losses associated with the impacts of the 1982-1983 El Niño event exceed $8 billion, and of these, U.S. losses exceeded $1.5 billion. During the 1982-1983 event, for example, California experienced, on average, 150 percent of normal rainfall with some areas receiving 200-300 percent of normal rainfall. We are predicting that such a scenario is much more likely than normal this winter in California. If this scenario plays out, Southern California is likely to experience the most severe impacts in the State. The confidence these forecasts will be correct is roughly 60 percent.

Our forecasts are based on using models to make a series of predictions using different atmospheric conditions. It is important to stress that these predictions show considerable variability amongst themselves with some showing near normal rainfall while others show rainfall 300% of normal. However, a majority of our predictions show significantly increased rainfall, similar to the 82-83 event and we are predicting that this is the most likely outcome. As is typical of an El Niño event, our predictions also show normal or above normal precipitation for much of the Southern States and dryer than normal conditions in the Northwest.

Given such a forecast, you might now ask what is being done to prepare the U.S. For this El Niño? Economic studies suggest that better climate predictions, if used appropriately, could reduce extreme seasonal climate damage costs in the U.S. by 25 percent, or $2.7 billion, annually in the agricultural sector alone. Water, energy, and transportation managers as well as farmers could plan and avoid or mitigate losses with accurate and timely predictions. However, there are more than economic studies to illustrate that a targeted program to improve U.S. regional climate forecasts and utilization of these forecasts to mitigate impacts of extreme events would be extremely valuable to the U.S.

During the 1980's, research on the El Niño-Southern Oscillation (ENSO) cycle of ocean-atmosphere interactions in the tropical Pacific produced strong insights into the predictability of ENSO and tantalizing evidence of the potential applications of this new forecasting ability in addressing societal issues related to agriculture, water resource management, and public health and safety. During the same period, some tropical countries, such as Australia, Brazil and Peru, began to demonstrate the value of incorporating ENSO-related forecast information into agricultural and resource management decisions.

For example, climate researchers at Scripps began to provide experimental climate forecasts 1-12 months in advance. Agricultural planners in Australia, Brazil and Peru used these forecasts to develop appropriate regional/sectoral guidance products advising farmers to adjust fertilizers, planting schedules and/or crop types to avoid crop failures and maximize yields. Their efforts demonstrate that use of seasonal climate forecasts have been key in protecting and enhancing production and profit margins) during periods of environmental stress (see Figure 2).

Another example which illustrates the value of climate forecasts occurred in Northeastern Brazil. In the early 1990's Northeastern Brazil was experiencing a period of extended drought. Some of our Brazilian colleagues inquired as to weather the forecasting and modeling efforts at SIO could help them to predict whether or not the drought would continue. In collaboration with Germany's Max Planck Institute, Scripps scientists predicted that the next expected rainy season (February, March, April, May) would fail to produce normal levels of precipitation. Under these projected drought conditions, the Brazilian City of Fortaleza (a city of several million people) would entirely deplete its water resources. To prevent this hardship the Brazilian government invested in a canal to Fortaleza. As predicted, the rains failed that season. Because they were given advance warning, the Brazilians finished the canal in time, saved millions of dollars in damages and protected public health.

Partially as a result of these early successes, the NOAA decided to work with other countries to establish an International Research Institute for Climate Prediction (IRI). Its mission is to provide experimental forecast guidances on seasonal-to-interannual time scales for use by affected communities around the world. Scripps and Columbia University's Lamont-Doherty Earth Observatory co-host the Institute. Working with an extensive network for research and applications centers around the world, the IRI will provide the necessary scientific and institutional focus for a multi-national "end-to-end" prediction program supporting the development and production of forecasts of changing physical conditions (e.g. temperature and rainfall) on year-to-year time scales, assessments of regional consequences of those variations, and the application of this information to support practical decision making in critical sectors like agriculture, water resources, fisheries, emergency preparedness and public health and safety.

Recently the IRI was asked by USAID to provide a forecast for East Africa so that USAID could make informed decisions about its programs in this area. The East African long rains failed in 1996 and Scripps predicted that the short rains were also likely to fail at the end of 1996. After they did fail, USAID, extremely concerned about impacts on the stability of the region if the 1997 long rains were to fail, asked scientists to prepare a consensus forecast on the 1997 long rains. Scientists agreed that the long rains were not likely to fail. As a result USAID was able to devote their limited resources to other areas.

Closer to home, Scripps is reaching out to affected communities on several levels. I have been working with the Southern California sport fishing industry (more than a billion dollar per year industry in California) to help it benefit from this year's predicted El Niño. During an El Niño, sport fishing is generally better than normal. Providing the sport fishing fleet with an El Niño forecast allows boat owners to increase advertising, make needed repairs to boats, and add crew as necessary. Partially as a result of the forecasts, the industry is enjoying a boom year.

Scripps is also engaged in a long term project to make the best use of climate forecasts to improve California's use of its scarce water resources. The goal of this research is improved forecasts of a range of hydrological elements from floods to multi-year fluctuations in the water supply. While the major focus is upon water resources in California, there are immediate applications to similar issues in the neighboring western states and other mountainous regions. For this year's event, Scripps is leading a team of University of California scientists to "down scale" global climate predictions to describe impacts on local water and energy supplies.

In an effort to reach out to potentially impacted communities, Scripps and the California Department of Boating and Waterways hosted a workshop on August 19, 1997, to describe and discuss the possible coastal impacts of severe storms that might be associated with an El Niño condition in the Pacific Ocean this coming winter. The event was attended by several hundred representatives from California cities and counties who were able to discuss preparedness measures that could mitigate projected hazards and costs. Scientists were also able to discuss ways to coordinate efforts to capture pre- and post-event environmental parameters to gauge the effects of a severe El Niño event on the near-shore, beach and back-beach environments. The goal of this research is to improve prediction of impacts of future events.

As a follow up to that meeting, Scripps researchers are working closely with potentially affected cities. These include Los Angeles, Long Beach and others throughout the region. We are encouraging the cities to prepare risk assessments and a list of possible mitigation measures. This information, coupled with the likelihood forecasts will be correct, will allow the cities to develop science-based cost/benefit analyses to decide what level of mitigation is appropriate. Potentially impacted cities are showing a great deal of interest and sophistication about the predicted El Niño. We believe this is due to NOAA's ongoing efforts to education the public about El Niño impacts and the evolving skill of forecasts.

Reducing losses and capitalizing on opportunities like the examples I have just described requires that predictions of El Niño-related extreme conditions be transformed into usable information which is effectively conveyed to Federal, State and local agencies, private industry and the public. Unfortunately, while there is an ongoing program in seasonal-to-interannual climate research and experimental forecasting, there is no program targeted at applying these new insights in a focused, mitigation research effort.

Efforts like those at SIO, and NOAA's National Weather Service effort to prepare western regional water managers for this event will mitigate some damage. These efforts, however, are at best ad hoc and rely on the arbitrary confluence of interested scientists and decision makers in affected regions. Given the magnitude of the predicted El Niño and its anticipated consequences, it is unfortunate that a more integrated hazard mitigation research effort does not exist.

In this context, I am encouraged by the initiative which today's hearing represents and, on behalf of Scripps, encourage Congress to call for and support the development of a focused, multi-agency national program to improve and utilize climate prediction data in weather-related hazard mitigation and disaster planning. To be most effective, such a program should be organized around the development of regional partnerships of scientists, government agencies and private sector interests, and focused on the use of emerging climate forecasting technology in critical sectors. It is particularly important for these regional partnerships to include emergency response officials and community planners at all levels of government.

Such a program will allow the US to build optimal decisionmaking into weather- and climate-sensitive sectors of the economy, such as agriculture, water-resource management, insurance, energy production and use, transportation, and construction. The nation has invested in both long-term climate research and experimental forecasting. It is critical the these investments now be used to protect life and property and ensure the greatest benefit for the U.S. economy.

Before closing, I want to state one final concern--the funding level recommended by the House for NOAA's Interannual and Seasonal Climate Research Program. I am concerned that the House Commerce/ State/Justice Appropriations bill does not include adequate funding for the research and observational programs necessary to predict El Ninos in the future.

The House Appropriations Committee did not fully fund the Administration's request for NOAA's Interannual and Seasonal Research Program. What is not funded is the $4.9M increase for the Interannual and Seasonal Research program which was to fund operations of the TOGA observing system. The $4.9 million is to ensure a stable funding base for the Tropical ocean-Global Atmosphere (TOGA) observing system. This initiative funds in-situ components which have already been shown to provide essential measurements for skillful forecasts of the El Nino phenomenon. These buoys measure, in real time, across the Equatorial Pacific temperature, currents, winds and other environmental variables. These measurements enable scientists to make predictions with a 6-12 month lead time. In fact, the TOGA observing system showed the successive propagation of the El Niño signal from the Western Pacific to South America (see Figure 3). These measurements, taken between December 1996 and February 1997, a time when the El Niño was not apparent at the ocean's surface, give us confidence in our long-range forecasts. It is critical that funds be included in the final appropriations bill to maintain our forecasting capability.

Continued US population growth and increased urbanization and concentration of capital and physical plant in hazard-prone areas virtually guarantee that economic losses from weather-related disasters will continue to rise. As the nation sacrifices to balance the federal budget while still investing in economic growth, health and national security, the costs of unmitigated natural disasters are no longer affordable. A dialogue must begin now to improve prediction, mitigation and response to natural disasters and reduce future costs.

Mr. Chairman, I thank you for the opportunity to appear before you today and I encourage you and your colleagues to continue to look for ways to support the use of the current predictions to mitigate costs of extreme weather events and capitalize on economic and strategic opportunities which these forecasts provide.

SST Anomaly Forecast for Dec/Jan/Feb 1997/1998
Made 27 Nov 1996

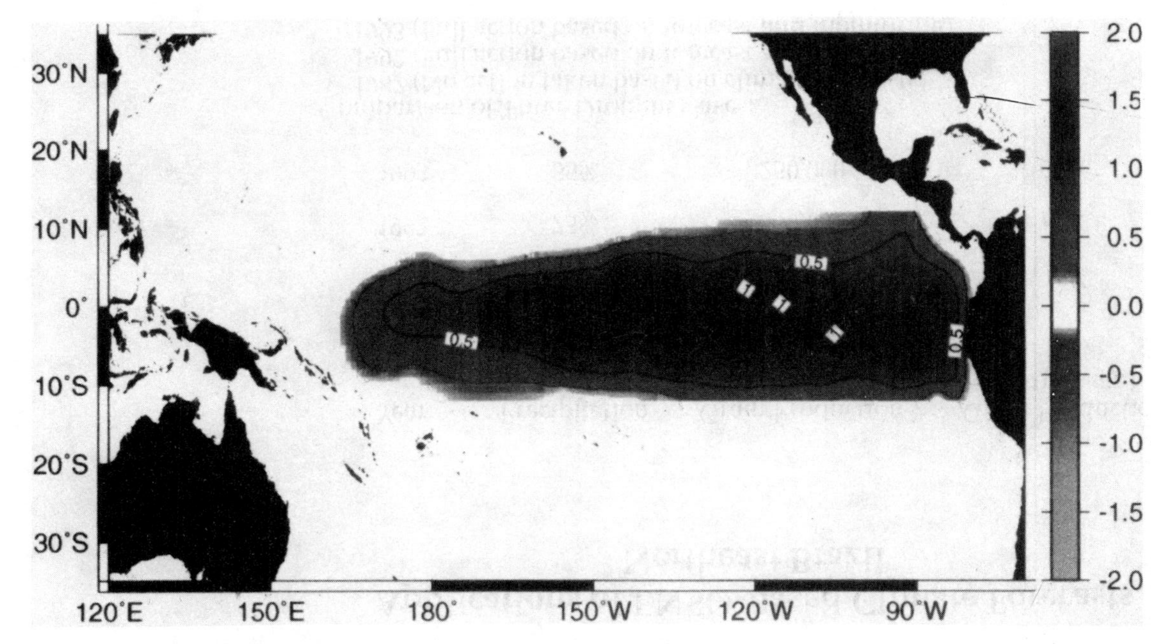

Scripps Institution of Oceanography /
Max-Plank-Institut fuer Meteorogie
HCM Version T3.0

Figure 2

Applications of ENSO-Based Climate Forecasts
Northeast Brazil

Year	Precipitation (% of mean)	Grain Production (in tons)	Grain Production (% of mean)
Mean		650,000	
1987	70%	100,000	15%
1992	73%	530,000	82%
1993	55%	250,000	38%

Comparison of Three Drought Cases:
1987 (No action taken based on climate forecasts)
1992 (Full action based on forecast and monitoring)
1993 (Full action based on forecast and monitoring)

Provided by: FUNCEME (Ceará's Foundation for Meteorology and Water Resources)

Figure 3

Monthly Mean TAO Temperature Anomalies (C)

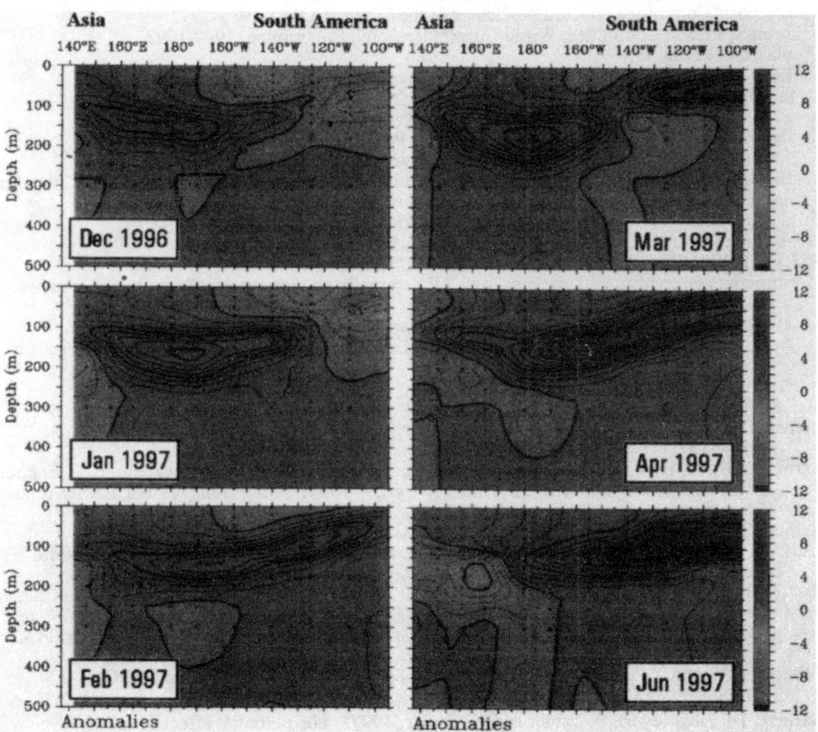

CURRICULUM VITA
TIM P. BARNETT

EDUCATION: B.A., Physics, Pomona College
M.S., Physical Oceanography,
Scripps Institution of Oceanography,
Ph.D., Physical Oceanography,
Scripps Institution of Oceanography

PROFESSIONAL July 1971 - present
EXPERIENCE: Research Marine Physicist
Academic Administrator (1971-1982)
Scripps Institution of Oceanography
University of California, San Diego
La Jolla, California, 92093-0224

SELECTED HONORS AND AWARDS
- Member, Committee on Climate Change in the Ocean, CCCO, 1989-1992
- Joint Scientific Committee, World Climate Research Programme, 1990-1992 via CCCO Executive
- Elected to International Association for the Physical Sciences of the Ocean (IAPSO) Committee on Tides and Mean Sea Level for three-year term (1983-1986).
- Expert Witness, Review of Global Warming, West German Bundestag, 3/90
- Elected, Councilor, 3 year term, American Meteorological Society, 1/92-1/95
- Elected Fellow, American Meteorological Society, Fall, 1991 (This is different than Councilor nomination above)
- Special Creativity Award, National Science Foundation, 1991 and 1992
- Awarded Sverdrup Gold Medal, (AMS), Fall, 1992

SELECTED OTHER PROFESSIONAL ACTIVITIES
- Numerous interviews with the press and TV nationally and internationally regarding weather, climate, El Nino, and greenhouse effects.
- Maintain scientific exchange program with Max Planck Institut, Hamburg, W. Germany
- Advise different branches of government (e.g., NOAA, DOE, NASA, JPL, EPA) on climate research and remote sensing.
- Over 200 talks to scientific and lay groups

SELECTED PUBLICATIONS

Barnett, T.P., E. Kirk, M. Latif, and E. Roeckner, 1991: On ENSO physics, *J. Climate*, **4**, 487-515.

Barnett, T.P., L. Bengtsson, K. Arpe, M. Flugel, N. Graham, M. Latif, J. Ritchie, E. Roeckner, U. Schlese, U. Schulzweida, and M. Tyree, 1994: Forecasting global ENSO-related climate anomalies, *Tellus*, **46A**, 381-397. .

Barnett, T.P., 1995: Monte Carlo climate forecasts. *J. Clim.*, **8(5)**, 1005-1022.

Latif, M. and T.P. Barnett, 1996: Decadal climate variability over the North Pacific and North America: Dynamics and predictability. *J. Clim.*, **9**, 2407-2423.

Santer, B.D., T.M.L. Wigley, T.P. Barnett, and E. Anyamba, 1996: Detection of Climate Change and Attribution of Causes. Chapter 8 in: *IPCC, 1996: Scientific Assessment of Climate Change, Intergovernmental Panel on Climate Change (IPCC),* Report Prepared for IPCC by Working Group I, WMO/UNEP, June, 1996.

Schneider, N. and T.P. Barnett, 1997: The Indonesian throughflow in a coupled GCM. *J. Geophys. Res.*, **102(C6)**, 12,341-12,358.

Barnett, T.P., G. Hegerl, B. Santer and K. Taylor, 1997: The potential effect of GCM uncertainties on greenhouse signal detection, *J. Clim.*, in press.

95-96

Scripps Institution of Oceanography
Financial Disclosure of NOAA+EPA fund
FY95-96

SPONSOR	AMOUNT	START	END	TITLE
NOAA	121551	950705	961029	THEORETICAL & NUMERICAL STUDY OF SURFACE WAVER SPECTRA TRANSFORMATION
NOAA	70000	951101	961031	SCIENTIFIC INFORMATION MANAGEMENT OF TEMPERATURE DEPTH OBSERVATIONS OVER THE PACIFIC OCEAN
NOAA	24950	951101	970930	SUPPORT FOR LONG TERM PLANNING OF NOAA ENVIROMENTAL DATA & INFORMATION MANAGEMENT
NOAA	6750	951106	960806	SEA GRANT STATE FELLOWSHIP
NOAA	50000	930701	970630	US GLOBEC: MODELLING STUDIES OF COUPLED BIOLOGICAL/PHYSICAL PROCESSES
NOAA	112036	950701	970630	IRICP RESEARCH & DEVELOPMENT PILOT PROJECT
NOAA	475000	950901	961031	GLOBAL DRIFTER MEASUREMENTS OF VELOCITY SST & SEA LEVEL PRESSURE
NOAA	93000	950901	960831	THE LAMONT/SCRIPPS CONSORTIUM FOR CLIMATE RESEARCH
NOAA	1812747	951101	960430	THE LAMONT/SCRIPPS CONSORTIUM FOR CLIMATE RESEARCH
NOAA	3410000	931001	960930	CALIFORNIA SEA GRANT COLLEGE
NOAA	50000	951001	960930	95-96 SABRE EGG SURVEY & DATA SYNTHESIS & INTEGRATION
NOAA	24000	931001	960930	CALIFORNIA SEA GRANT COLLEGE
NOAA	78975	951201	961130	CALIFORNIA SEA GRANT COLLEGE
NOAA	80000	951001	970930	DEFICIENT GENETIC DIVERSITY IN TRANSPLANTED EELGRASS
NOAA	40000	951001	960930	COASTAL OCEAN SCIENCE & EDUCATION
NOAA	72000	960201	970131	NATIONAL SEA GRANT FELLOWSHIPS
NOAA	69890	951101	961031	PHYTOPLANKTON & PRIMARY PRODUCTIVITY STUDIES IN SUPPORT OF THE US ALMR PROGRAM
NOAA	11080	960501	970430	DECADAL HYDROCLIMATIC VARIABILITY IN WESTERN NORTH AMERICA
NOAA	19900	960401	970331	HABITAT & TROPHIC LINKAGES IN SALT MARSHES: GEOGRAPHIC COMPARISONS
NOAA	180659	960401	970331	A STUDY OF THE ROLE OF THE INDIAN OCEAN IN THE GLOBAL CLIMATE SYSTEM
NOAA	141160	960401	970331	AIR SEA EXCHANGE PROCESSES IN TOGA COARE
NOAA	770000	960401	970331	GLOBAL DRIFTER MEASUREMENTS OF VELOCITY SST & PA
NOAA	24000	960701	970630	REGULATION OF GENE EXPRESSION IN RHEUMATOID ARTHRITIS
NOAA	55000	960501	970430	AUTOMATED IMPROVED ESTIMATES OF CLOUD COVER OF LAND AREAL EXTENT OF SNOW COVER
NOAA	2198108	960501	961231	THE LAMONT/SCRIPPS CONSORTIUM FOR CLIMATE RESEARCH

96-97

Wed Sep 3 08:22:27 1997   1

Scripps Institution of Oceanography
Financial Disclosure of NOAA & EPA funds
FY96-97

SPONSOR	AMOUNT	START	END	TITLE
NOAA	27631	960701	970630	JIMO TASK IV INTEGRATED USE OF GOES GVAR & NOAA POLAR ORBITER SATELLITE DATA
NOAA	28334	960601	970531	INTEGRATION OF OCEAN MODELING OVER THE PACIFIC BASIN
NOAA	56666	960601	970531	INTEGRATION OF OCEAN MODELING OVER THE PACIFIC BASIN
NOAA	40000	960701	970630	ACCURATE DETECTION OF PATTERNS OF SPATIAL & TEMPORAL CHANGE IN SATELLITE MODELS
NOAA	41000	960801	970731	DIAGNOSTIC & MODELING STUDY OF THE GCIP ENERGY & MOISTURE BUDGETS
NOAA	43111	960701	970630	THE ROLE OF LIFE HISTORIES IN FAUNAL RECOVERY OF SO CALIF RESTORED WETLANDS
NOAA	9000	960901	970831	TROPHIC LINKAGES IN CREATED & NATURAL SALT MARSHES
NOAA	500000	960901	970831	SCRIPPS EXPERIMENTAL CLIMATE PREDICTION CENTER
NOAA	17140	960901	970831	EXTENDING THE COMPREHENSIVE OCEAN ATMOSPHERE DATA SET TO THE UPPER OCEAN & THE TROPOSPHERE
NOAA	190000	960901	970831	NATIONAL WEATHER SERVICE WEST COAST DRIFTER PROGRAM
NOAA	40000	961001	970930	FURTHER DEVELOPMENT & USE OF A RAPID SURVEY METHOD FOR FISH EGGS
NOAA	17000	960901	971130	MONITORING & DEVELOPMENT OF BIOPHYSICAL INDICES OF THE SE BERING SEA
NOAA	2697000	961001	970930	CALIFORNIA SEA GRANT COLLEGE
NOAA	2730146	961001	990930	CALIFORNIA SEA GRANT COLLEGE
NOAA	19991	960901	970831	UNDERWAY CHEMICAL INSTRUMENTATION FOR THE MEASUREMENT OF CO2 SYSTEM PARAMETERS
NOAA	30000	960901	961130	SCIENTIFIC INFORMATION MANAGEMENT OF TEMPERATURE DEPTH OBSERVATIONS OVER THE PACIFIC OCEAN
NOAA	-88286	960705	961029	THEORETICAL & NUMERICAL STUDY OF SURFACEW WAVER SPECTRA TRANSFORMATION ON FLOW CAUSED BY UNDERW
NOAA	73364	960901	971031	PHYTOPLANKTON & PRIMARY PRODUCTIVITY STUDIES IN SUPPORT OF THE US ALMR PROGRAM
NOAA	49745	970201	980131	JIMO TASK II OFF-CAMPUS SATELLITE DATA PROCESSING
NOAA	4089190	970101	971231	THE LAMONT/SIO CONSORTIUM FOR CLIMATE
NOAA	36000	970201	980131	CALIFORNIA SEA GRANT COLLEGE
NOAA	75000	970315	980314	MAINTENANCE & TRANSFER OF THE INTERNATIONAL CALIBRATION SCALE FOR CARBON DIOXIDE
NOAA	6750	970115	971014	CALIFORNIA SEA GRANT COLLEGE STATE FELLOWSHIP
NOAA	3638000	970301	990228	CALIFORNIA SEA GRANT COLLEGE
NOAA	47818	970701	980331	CALIFORNIA SEA GRANT COLLEGE
NOAA	655000	970401	980331	GLOBAL DRIFTER MEASUREMENTS OF VELOCITY SST & PA
NOAA	11364	970501	980430	FURTHER DEVELOPMENT & USE OF A RAPID SURVEY METHOD FOR FISH EGGS
NOAA	550156	970401	970531	THE LAMONT/SIO CONSORTIUM FOR CLIMATE RESEARCH
NOAA	59609	970501	980430	JIMO: PROCESS STUDIES OF PHYSICAL-BIOLOGICAL INTERACTIONS & LARVAL FISH ON GEORGES BANK
NOAA	88400	970601	981130	SPORT FISH IN CALIFORNIA WATERS

97-98

Wed Sep 3 08:22:54 1997   1

Scripps Institution of Oceanography
Financial Disclosure of NOAA/EPA
funds
FY97-98

SPONSOR	AMOUNT	START	END	TITLE
NOAA	28531	970701	980630	JIMO TASK IV-INTEGRATED USE OF GOES GVAR & NOAA POLAR ORBITER SATELLITE DATA IN OPERATIONAL WEAT
NOAA	40000	960701	980630	ACCURATE DETECTION OF PATTERNS OF SPATIAL & TEMPORAL CHANGE IN SATELLITE MODEL & IN SITU DATA US
NOAA	24000	970701	980630	THE REGULATION OF GENE EXPRESSION IN RHEUMATOID ARTHRITIS

Chairman CALVERT. Thank you, Doctor, for your testimony.
Next, Dr. Solow, you may begin your testimony.

TESTIMONY OF ANDREW R. SOLOW, DIRECTOR, MARINE POLICY CENTER, WOODS HOLE OCEANOGRAPHIC INSTITUTE, WOODS HOLE, MA

Mr. SOLOW. Thank you, Mr. Chairman.

I am Andrew Solow. I am an associate scientist at the Woods Hole Oceanographic Institution and the Director of the Institution's Marine Policy Center. My remarks will focus on the economic value of El Niño prediction.

As you have heard, the El Niño Southern Oscillation, or ENSO, is a natural quasi-periodic redistribution of heat in the equatorial Pacific Ocean. In broad terms, ENSO can be characterized as exhibiting one of three phases: warm, cold, and neutral. These phases, in turn, can be characterized by the distribution of sea surface temperatures in the Pacific. During the warm phase, the eastern equatorial Pacific is anomalously warm. The occurrence of this phase is commonly referred to as an El Niño event. During the cold phase, the eastern equatorial Pacific is anomalously cool. Finally, ENSO can exhibit a neutral phase during which the eastern equatorial Pacific is neither anomalously warm nor anomalously cold.

Since 1877, there have been around 30 El Niño events and 22 cold events. Thus, on average, an El Niño event occurs every 4 years. While the frequency of events fluctuates around this average, there is no evidence of a systematic change in the frequency over the historical period.

Through its effects on atmospheric circulation, ENSO has a significant effect on regional climate on both sides of the Pacific. In many parts of the world, ENSO is the largest source of climate variability on the time scale of 1 to 10 years. In the United States, the clearest ENSO signal occurs in the southeast, where El Niño events are associated with cool, wet winters and warm, dry summers. El Niño events are also associated with reduced tropical storm activity in the Atlantic basin. Precipitation over the western United States is also affected.

It has recently been suggested that some of the variability in tornado activity is related to ENSO, although further work is needed to establish this relationship.

The ability to predict ENSO phase up to a year in advance has improved over the past 20 years or so. Because the phase of ENSO has an effect on regional climate, this has also led to improved skill in long-range climate prediction. This has raised a question about the economic value of ENSO prediction and long-range climate prediction generally. Beyond its academic interest, the answer to this question would be useful in determining the appropriate level of investment in maintaining and improving ENSO prediction.

The value of ENSO prediction is defined as the expected increase in benefits to society arising from the use of the prediction in economic decisionmaking. It is important to emphasize that this is smaller than the total value of ENSO-related losses. The reason is that, even with perfect foresight, not all of these losses can be avoided. It is also important to emphasize that this value is associated with the repeated use of the prediction over time. It is certainly possible that the prediction in any particular year will be incorrect and that its use will lead to bad decisions being made. However, over time, this will be more than offset by gains from using correct predictions.

Finally, while it may seem that ENSO prediction has substantial value to the financial sector—insurance and commodities markets—the actual value is probably small. The reason is that, to a first approximation, these sectors involve transfers from one part of society to another and not genuine productive activity.

With Richard Adams of Oregon State University, Rodney Weiher of NOAA, and others, I have conducted studies of the value of ENSO prediction to certain economic activities in the United States. The most comprehensive of these studies focused on U.S. agriculture.

The use of ENSO prediction in agriculture is analogous to an improvement in agricultural technology. It results in a greater quantity of agricultural products being supplied to the market at a lower price. The study estimated the value of perfect ENSO phase prediction to producers and consumers of U.S. agricultural products at around $300 to $400 million per year, or around 1 percent of the value of annual agricultural production. Applying a discount rate of 5 percent, this amounts to around $3 billion over a 10-year period.

In contrast, the agricultural losses due to a single El Niño event are around $2 billion. However, as I noted, even with perfect foresight, it would not be possible to avoid all these losses.

Improved ENSO prediction can also have indirect value. For example, there has been considerable concern recently about oxygen depletion or hypoxia in part of the Gulf of Mexico—what has come to be called "The Dead Zone." This hypoxia is caused in large part by the delivery of nutrients to the Gulf by the Mississippi River. The source of some of these nutrients is over-fertilization by farmers in the Mississippi Basin. Over-fertilization is a rational response to uncertainty about weather. By reducing this uncertainty, improved ENSO prediction would reduce the incentive to over-fertilize, thereby reducing nutrient-loading in the Gulf and ameliorating the hypoxia.

By way of a bottom line, I would estimate conservatively the total value of perfect ENSO phase prediction to the U.S. economy at around $1 to $2 billion per year. From the estimates that I have seen of the magnitude of public investments in maintaining and improving ENSO prediction, this return is very high. Incidentally, the potential value of ENSO prediction in other parts of the world is almost certainly greater than in the United States and the provision of predictions could be viewed as a form of foreign aid.

The results that I have given pertain to the prediction of ENSO phase. The proportion of climate variability explained by ENSO phase alone remains rather modest. Even perfect ENSO phase prediction is far from perfect climate prediction. One area for further scientific work is the prediction of the magnitude and other aspects of ENSO events. This would also contribute to improved skill in long-range prediction and would have additional value to society.

Finally, the question commonly arises: If ENSO prediction has such value, then why should not the private sector pay for it?

There is an answer to this question that even an economist like Milton Friedman—who is no fan of big government—would accept. Economic theory says that the socially optimal price of an ENSO prediction equals the marginal cost of supplying it—that is, the cost of supplying it to one additional user. Clearly, this marginal cost is zero. Thus, to make ENSO prediction anything but free is not socially optimal. However, it is not possible to earn a fair rate of profit providing a free prediction. Even if an attempt were made to charge for the prediction, it would be impossible to prevent its free dissemination.

Clearly, no private firm will undertake to provide a free prediction. This is an example of a public good and it is a legitimate role of government to provide public goods to society.

Thank you very much.

[The prepared statement of Mr. Solow follows:]

Subcommittee on Energy and Environment

U.S. House of Representatives

For the hearing record of September 11, 1997

Statement by

Dr. Andrew R. Solow

Woods Hole Oceanographic Institution

<u>The El Niño-Southern Oscillation</u>

The **El Niño**-Southern Oscillation or **ENSO** refers to a natural, quasi-periodic redistribution of heat and momentum in the equatorial Pacific Ocean. In broad terms, ENSO can be characterized as exhibiting one of three phases: warm, cold, and neutral. These phases can, in turn, be characterized by the distribution of sea surface temperatures in the equatorial Pacific. During the warm phase, the eastern equatorial Pacific is anomalously warm. The occurrence of this phase is commonly referred to as an El Niño event. During the cold phase, the eastern equatorial Pacific is anomalously cool. This is variously referred to as a La Niña or El Viejo event. Finally, ENSO can exhibit a neutral phase, in which the eastern equatorial Pacific is neither anomalously warm nor anomalously cold.

Since 1877, there have been around 30 El Niño events and 22 cold events. Thus, on average, an El Niño event occurs every four years. While the frequency of events fluctuates around this average, there is no evidence of a systematic change in the frequency over the historical period.

Through its effects on atmospheric circulation, ENSO has a significant effect on regional climate on both sides of the Pacific. In many parts of the world, ENSO is the largest source of climate variability on the time scale of 1-10 years. In the U.S., the clearest ENSO signal occurs in the southeast, where El Niño events are associated with cool, wet winters and warm, dry summers.

El Niño events are also associated with reduced tropical storm activity in the Atlantic basin. Precipitation over the western U.S. may also be affected by ENSO, although the signal is less clear. It has recently been suggested that some of the variability in tornado activity is related to ENSO, although further work is needed to establish this relationship.

The value of ENSO prediction

The ability to predict ENSO phase up to a year in advance has improved over the past twenty years or so. Because the phase of ENSO has an effect on regional climate, this has also led to improved skill in long-range climate prediction. This has raised a question about the economic value of ENSO prediction and long-range climate prediction generally. Beyond its academic interest, the answer to this question would be useful in determining the appropriate level of investment in maintaining and improving ENSO prediction.

The value of ENSO prediction is defined as the expected increase in benefits to society arising from the use of the prediction in economic decisionmaking. It is important to emphasize that this is smaller than the total value of ENSO-related losses. The reason is that, even with perfect foresight, not all of these losses can be avoided. It is also important to emphasize that this value is associated with the repeated use of the prediction over time. It is certainly possible that the prediction in any particular year will be incorrect and that its use will lead to bad decisions being made. However, over time, this will be more than offset by gains from using correct predictions. Finally, while it may seem that ENSO prediction has substantial value to the insurance sector, commodities markets, etc., the actual value is probably small. The reason is that, to a first approximation, these sectors involve transfers from one part of society to another and not genuine productive activity.

With Richard Adams of Oregon State University, Rodney Weiher of NOAA, and others, I have conducted studies of the value of ENSO prediction to certain economic activities in the U.S. These are the first rigorous studies of their kind. The most comprehensive of these studies focused on U.S. agriculture. The use of ENSO prediction in agriculture is analogous to an improvement in agricultural technology: it results in a greater quantity of agricultural products being supplied to the market at a lower price. The study estimated the value of perfect ENSO phase prediction to producers and consumers of U.S. agricultural products at around $300-400 million per year, or around 1% of the value of annual agricultural production. Applying a discount rate of 5%, this amounts to around $3 billion over a 10-year period. In contrast, the agricultural losses due to a single El Niño event are around $2 billion. However, as I noted, even with perfect foresight, it would not be possible to avoid all of these losses.

In a second study, we estimated the annual value of perfect

ENSO prediction to the Pacific coho salmon fishery at around $25 million. This represents around 1-2% of the annual value of the fishery. We are currently developing a study of the value of ENSO prediction to the hydroelectric power generating industry. If the level of interest of the industry is any guide, we again expect to find a significant value.

Improved ENSO prediction can also have indirect value. For example, there has been considerable concern recently about oxygen depletion or hypoxia in part of the Gulf of Mexico - what has come to be called the Dead Zone. This hypoxia is caused in large part by the delivery of nutrients to the Gulf by the Mississippi River. The source of some of these nutrients is over-fertilization by farmers in the Mississippi Basin. Over-fertilization is a rational response to uncertainty about weather. By reducing this uncertainty, improved ENSO prediction would reduce the incentive to over-fertilize, thereby reducing nutrient-loading in the Gulf of Mexico and ameliorating the hypoxia.

Conclusion

By way of a bottom line, I would estimate conservatively the total value of perfect ENSO prediction to the U.S. economy at around $1-2 billion per year. From the estimates that I have seen of the magnitude of public investments in maintaining and improving ENSO prediction, this return is very high. Incidentally, the potential value of ENSO prediction in other parts of the world is almost certainly greater than in the U.S. and the provision of predictions could be viewed as a form of foreign aid.

The results that I have given pertain to the prediction of ENSO phase. The proportion of climate variability explained by ENSO phase alone remains rather modest. Even perfect ENSO phase prediction is far from perfect climate prediction. One area for further scientific work is the prediction of the magnitude and other aspects of ENSO events. This would also contribute to improved skill in long-range climate prediction and would have additional value to society.

Finally, the question commonly arises: If ENSO prediction has such value, then why should not the private sector pay for it? There is an answer to this question that even economists like Milton Friedman, who is no fan of big government, would accept. Economic theory says that the socially optimal price of an ENSO prediction equals the marginal cost of supplying it - that is, to the cost of supplying it to one more user. Clearly, this marginal cost is zero. Thus, to make ENSO prediction anything but free is not socially optimal. However, it is not possible to earn a fair rate of profit providing a free prediction. Even if an attempt was made to charge for the prediction, it would be impossible to prevent its free dissemination. This is an example of a public good and it is a legitimate role of government to provide public goods to society.

ANDREW R. SOLOW
Marine Policy Center
Woods Hole Oceanographic Institution
Woods Hole, MA   02543
(508) 457-2746

Experience

1995-present	Director, Marine Policy Center Woods Hole Oceanographic Institution
1987-present	Assistant Scientist, Associate Scientist Woods Hole Oceanographic Institution
1985-1987	Post-doctoral Research Fellow Woods Hole Oceanographic Institution

Education

1986	PhD, Statistics Stanford University
1977	BA, Economics Harvard University

Memberships

EPA National Advisory Committee on Environmental Policy and Technology

Scientific Working Group, Intergovernmental Panel on Climate Change

Editorial Board, Environmental and Ecological Statistics

Editorial Council, Journal of Environmental Economics and Management

Editorial Board, Ecology

Research interests

environmental statistics; statistical ecology; spatial statistics; time series analysis; Bayesian statistics

Publications

Solow, A.R. 1985. Bootstrapping correlated data. <u>Mathematical Geology</u>, 17, 769-775.

Solow, A.R. 1986. Mapping by simple indicator kriging. <u>Mathematical Geology</u>, 18: 335-352.

Solow, A.R. and Gorelick, S.M. 1986. Estimating monthly streamflow values by cokriging. <u>Mathematical Geology</u>, 18: 785-809.

Solow, A.R. 1987. The application of eigenanalysis to tide-gauge records of relative sea level. <u>Continental Shelf Research</u>, 7(6): 629-641.

Solow, A.R. 1987. Testing for climate change: An application of the two-phase regression model. <u>Journal of Climate and Applied Meteorology</u>, 26(10): 1401-1405.

Solow, A.R. 1988. Detecting changes through time in the variance of a long-term hemispheric temperature record: An application of robust locally weighted regression. <u>Journal of Climate</u>, 1(3): 290-296.

Solow, A.R. 1988. A Bayesian approach to statistical inference about climate change. <u>Journal of Climate</u>, 1(5): 512-521.

Solow, A.R. 1988 (1986). On assessing dry probabilities in offshore oil and gas exploration: An application of Bayes Theorem. In: Quantitative Analysis of Mineral and Energy Resources, Geol. Surv. Can., Ontario, Canada (C.F. Chung, A.G. Fabbri, and R. Sinding-Larsen, eds.), <u>NATO-Advanced Study Institutes Series C, Mathematical and Physical Sciences</u>, 223: 187-198.

Solow, A.R. and Broadus, J.M. 1988. A simple model of over-forecasting. <u>Monthly Weather Review</u>, 116(6): 1371-1373.

Solow, A.R., Bullister, J.L., and Nevison, C. 1988. An application of circular-linear correlation analysis to the relationship between Freon concentration and wind direction in Woods Hole, Massachusetts. <u>Environmental Monitoring and Assessment</u>, 10: 219-228.

Solow, A.R. and Broadus, J.M., 1989. Climate Catastrophe: on the Horizon or Not?, <u>Oceanus</u>, 32(2): 61-64.

Solow, A.R., 1989. Is it getting stuffy in here, or is it just my imagination? <u>Chance</u>, 2: 40-46.

Solow, A.R. 1989. Statistical modelling of storm counts. <u>Journal of Climate</u>, 2(2): 131-136.

Solow, A.R. 1989. Reconstructing a partially observed record of tropical cyclone counts. <u>Journal of Climate</u>, 2(11): 1253-1257.

Solow, A .R. 1989. Bootstrapping sparsely sampled spatial point patterns. <u>Ecology</u>, 70(2): 379-382.

Solow, A.R. 1989. A randomization test for independence of animal locations. <u>Ecology</u>, 70(5): 1546-1549.

Solow, A.R. 1989. A counter-intuitive result for mixture random fields. <u>Mathematical Geology</u>, 21: 379-382.

Solow, A.R. and Broadus, J.M. 1989. On the detection of greenhouse warming. <u>Climatic Change</u>, 15: 449-453.

Solow, A.R. and Broadus, J.M. 1989. Loss functions in estimating offshore oil resources. <u>Resources and Energy</u>, 11: 81-92.

Solow, A.R. 1990. Testing for density dependence: A cautionary note. <u>Oecologia</u>, 83(1): 47-49.

Solow, A.R. 1990. Geostatistical cross-validation; a cautionary note. <u>Mathematical Geology</u>, 22(6): 637-639.

Solow, A.R. 1990. Discriminating between models: An application to relative sea level at Brest. <u>Journal of Climate</u>, 3: 792-796.

Solow, A.R. 1990. A method for approximating multivariate normal orthant probabilities. <u>Journal of Statistical Computation and Simulation</u>, 37: 225-229.

Solow, A.R. 1990. A randomization test for misclassification probability in discriminant analysis. <u>Ecology</u>, 71(6): 2379-2382.

Solow, A.R. 1990. A note on the statistical properties of animal locations. <u>Journal of Mathematical Biology</u>, 29: 189-193.

Solow, A.R. and Broadus, J.M., 1990. Global Warming: Quo Vadis? <u>The Fletcher Forum of World Affairs</u>, 14(2): 262-267.

Solow, A.R. and Gallager, S.M. 1990. Analysis of capture efficiency in suspension feeding: application of nonparametric binary regression. <u>Marine Biology</u>, 107(2): 341-344.

Solow, A.R. and Nicholls, N. 1990. On the relationship between the Southern Oscillation and tropical cyclone frequency in the Australian region. <u>Journal of Climate</u>, 3: 1097-1101.

Solow, A.R. and Tyack, P. 1990. Inhomogeneity and apparent organization in animal behavior. <u>Biometrics</u>, 46: 837-839.

Solow, A.R. and Steele, J.H. 1990. On sample size, statistical power, and the detection of density dependence. <u>Journal of Animal Ecology</u>, 59: 1073-1076.

Cowan, D.F., Atema, J., and Solow, A.R. 1991. Moult staggering in the American lobster: A statistical analysis. _Animal Behavior_, 42: 863-864.

Gornitz, V. and Solow, A.R. 1991. Observations of long-term tide-gauge records for indications of accelerated sea level rise. In: Greenhouse-Gas-Induced Climatic Change: A Critical Appraisal of Simulations and Observations (M.E. Schlesinger, ed.), Elsevier Science, Amsterdam, 347-367.

Haney, J.C. and Solow, A.R. 1991. Analyzing quantitative relationships between seabirds and marine resource patches. _Current Ornithology_, 9: 105-162.

Solow, A.R. 1991. Is there a global warming problem? In: Global Warming: Economic Policy Responses (R. Dornbusch and J. Poterba, eds.), MIT Press, Cambridge, 7-28.

Solow, A.R. 1991. The nonparametric analysis of point process data: The freezing history of Lake Konstanz. _Journal of Climate_, 4(1): 116-119.

Solow, A.R. 1991. An exploratory analysis of the occurrence of explosive volcanism in the Northern Hemisphere, 1851-1985. _Journal of the American Statistical Association_, 86(413): 49-54.

Solow, A.R. 1991. On the statistical comparison of climate model output and climate data. In: Greenhouse-Gas-Induced Climatic Change: A Critical Appraisal of Simulations and Observations (M.E. Schlesinger, ed.), Elsevier Science, Amsterdam, 505-510.

Solow, A.R. 1991. A statistician's guide to global warming. _Proceedings of the American Statistical Association on Statistics and the Environment_, 14-21.

Solow, A.R. and Smith, W.K. 1991. Detecting cluster in a heterogeneous community sampled by quadrats. _Biometrics_, 47: 311-317.

Davis, C.S., Gallager, S.M., and Solow, A.R. 1992. Microaggregations of oceanic plankton observed by towed video microscopy. _Science_, 257(5067): 230-232.

Haney, J.C. and Solow, A.R. 1992. Testing for resource use and selection by marine birds. _Journal of Field Ornithology_, 63(1): 43-52.

Solow, A.R. and Patwardhan, A. 1992. On the consistency of the historic temperature record with greenhouse warming. _Environmetrics_, 3(4): 361-368.

Solow, A.R. 1992. Model-checking in non-stationary Poisson processes. _Applied Stochastic Models and Data Analysis_, 8: 129-132.

Solow, A.R. 1992. The response of sea level to global warming.
In: The World at Risk: _Natural Hazards and Climate Change_ (R.L.
Bras, ed.), American Institute of Physics, 38-42.

Solow, A.R. 1992. Detecting change in the composition of a multi-
species community. _Biometrics_, 50(2): 556-565.

Solow, A.R. and Gaines, A.G. 1993. Mapping water quality by local
scoring. _Canadian Journal of Statistics_, 21(2): 123-130.

Polasky, S., Solow, A.R., and Broadus, J.M. 1993. Searching for
uncertain benefits and the conservation of biological diversity.
Environmental and Resource Economics, 3: 171-181.

Solow, A.R., Polasky, S., and Broadus, J.M. 1993. On the
measurement of biological diversity. _Journal of Environmental
Economics and Management_, 24: 60-68.

Solow, A.R. and Starczak, V.E. 1993. Statistical analysis with an
ordered categorical regressor: The effect of rainfall on
attendance. _Water Resources Research_, 29(6): 1561-1564.

Solow, A.R. 1993. A simple test for change in community
structure. _Journal of Animal Ecology_, 62: 191-193.

Solow, A.R. 1993. Inferring extinction from sighting data,
Ecology, 74: 962-964.

Solow, A.R. 1993. Inferring extinction in a declining population,
Journal of Mathematical Biology, 32(1): 79-82.

Solow, A.R. 1993. On the efficiency of the indicator approach in
geostatistics. _Mathematical Geology_, 25(1): 53-57.

Solow, A.R. 1993. Spectral estimation by variable span log
periodogram smoothing: An application to annual lynx numbers.
Biometrical Journal, 35(5): 627-633.

Solow, A.R. 1993. Estimating record inclusion probability.
The American Statistician, 47(3): 206-208.

Solow, A.R. 1993. Comment on Professor Journel's paper.
Journal of Environmental and Ecological Statistics.

Kite-Powell, H. and Solow, A.R. 1994. A Bayesian approach to
estimating benefits of improved forecasts. _Meteorological
Applications_, 1: 351-354.

Smith, W. and Solow, A.R. 1994. An exact McNemar test for paired
binary Markov chains. _Biometrics_, in press

Solow, A.R., Tyack, P., and Sayigh, L. 1994. The statistical
analysis of whistle exchanges in dolphins. _Behavioral Processes_,
in press.

Solow, A.R. and Helser, T. 1994. Detecting extinction in sighting data. In: <u>Quantitative Methods in Conservation</u>, Springer-Verlag, in press.

Solow, A.R. and Patwardhan, A. 1994. Some model-based inference about global warming. <u>Environmetrics</u>, 5: 273-279.

Solow, A.R. and Polasky, S. 1994. Measuring biological diversity. <u>Journal of Ecological and Environmental Statistics</u>, 1: 95-107.

Solow, A.R. and Ratick, S.J. 1994. Conditional simulation and the value of information In: Geostatistics for the next Century, R. Simitrakopoulos (ed.), Kluwer Academic Publishers, 209-217.

Solow, A.R. 1994. Statistical methods in atmospheric science. In: Handbook of Statistics (G.P. Patil and C.R. Rao, eds.), Elsevier Science, 717-734.

Solow, A.R. 1994. Saturday effects in tanker oil spills: Comment. <u>Journal of Environmental Economics and Management</u>, 26(3): 293-296.

Solow, A.R. 1994. Comment on the National Research Council Report of Statistics and Physical Oceanography. <u>Statistical Science</u>, 9(2): 213-215.

Solow, A.R., 1994. On the Bayesian estimation of the number of species in a community, <u>Ecology</u>, 75(7): 2139-2142.

Solow, A.R. 1994. Estimating the size of a source population from a sample of matched parts. <u>Mathematical Geology</u>, in press.

Adams, R., Bryant, K., McCarl, B., Legler, D., O'Brien, J., Solow, A.R., and Weiher, R., 1995. Value of improved long-range weather information. <u>Contemporary Economic Policy</u>, 13: 10-19.

Polasky, S. and Solow, A.R. 1995. On the value of a collection of species. <u>Journal of Environmental Economics and Management</u>, 29: 298-303.

Smith, W., Solow, A.R., and Preston, P. 1995. An estimator of species overlap using a modified beta-binomial model. <u>Biometrics</u>, in press.

Solow, A.R. and Broadus, J.M., 1995. Issues in the measurement of biological diversity. <u>Vanderbilt Journal of Transnational Law</u>, 28(4): 695-702.

Solow, A.R. and Faber, D. 1995. A test of Przibram's Rule. <u>Journal of Zoology, Lond.</u>, 237(1): 159-161.

Solow, A.R. and Gaines, A.G. 1995. An empirical Bayes approach to monitoring water quality. <u>Environmetrics</u>, 6: 1-5.

Solow, A.R. and Patwardhan, A. 1995. Extracting a smooth trend from a time series: A modification of singular spectrum analysis. Journal of Climate, in press.

Solow, A.R., Smith, W. and Recchia, C. 1995. A conditional test of independence of Markov chains. Biometrical Journal, 8: 973-977.

Solow, A.R. and Steele, J.H. 1995. Scales of plankton patchiness: Biomass versus demography. Journal of Plankton Research, 17(8): 1669-1677.

Solow, A.R., Mound, L.A., and Gaston, K.F. 1995. Estimating the rate of synonymy. Systematic Biology, 44, 93-96.

Solow, A.R. 1995. Fitting population models to time series data. In: Ecological Time Series (T. Powell and J. Steele, eds.). Chapman and Hall, 20-27.

Solow, A.R. 1995. Testing for change in the frequency of El Niño events. Journal of Climate, 18(2), 2563-2566.

Solow, A.R. 1995. The goodness of fit of the cascade model. Ecology, in press.

Solow, A.R., 1995. Identifying regional seismic relationships by cross-correlation. Pageoph, 144(2): 301-306.

Solow, A.R., 1995. An exploratory analysis of a record of El Niño events, 1800-1987. Journal of American Statistical Assoc., 90(429): 72-77.

Solow, A.R., 1995. Estimating biodiversity: Calculating unseen species richness, Oceanus, 38(2): 9-10.

Solow, A.R. 1996. A test for a common upper endpoint in fossil taxa. Paleobiology, 22:406-410.

Solow, A.R., 1996. On fitting a population model in the presence of measurement error. Ecology, in press.

Solow, A.R., 1996. An empirical Bayes analysis of volcanic eruptions. J. Volcanology and Geothermal Res., submitted.

Solow, A.R., 1996. Bayesian methods about extinction times. Proceedings of the American Statistical Association, in press.

Solow, A.R., Curran, M.C., and Bentivegna, F. (1996). Estimating the order of a sequence of physiological changes from incomplete data. Biometrical Journal, in press.

Solow, A.R. and Sherman, K. 1996. Testing the stability of a predator-prey system. Ecology, in press.

Solow, A.R., Adams, R.F., Bryant, K.J., Legler, D.M., O'Brien, J.J., McCarl, B.A., Nayda, W., and Weiher, R. (1996). The value of improved ENSO prediction to U.S. agriculture. Climatic Change, in press.

Solow, A.R. and Smith, W.K. 1996. Estimation from an incomplete fossil record. Paleobiology, in press.

Solow, A.R. and Beet, A. 1997. On lumping in food webs. Ecology, in press.

Solow, A.R. and Smith, W.K. 1997. How surprising is a new record? American Statistician, submitted.

Solow, A.R. 1997. Dating the origin of cave art in southern Europe. Applied Statistics, submitted.

Kraus, S., Read, A., Solow, A.R., Baldwin, K., Spradlin, T., Anderson, E., and Williamson, J. 1997. Acoustic alarms reduce porpoise mortality. Nature, 388, 525.

Book reviews

Analytical Population Dynamics by S. Royama, Ecology.

Statistics for the Environment, 1994, edited by V. Barnett and K.F. Turkman, Journal of Applied Statistics.

Atmospheric Data Analysis by R. Daley, Journal of the American Statistical Association.

Design and Analysis of Ecological Experiments, 1994, edited by S. Scheiner and J. Gurevitch, Ecology.

Time Series Prediction, 1994, edited by A.S. Weigend and N.A. Gershenfeld, Science.

Statistical Data Analysis for Ocean and Atmospheric Sciences by H.J. Thiebaux, American Statistician.

Singular Spectrum Analysis by J. Elsner and A. Tsonis, Bulletin of the American Meteorological Society

Financial disclosure

Dr. Solow has received a total of approximately $48,000 over the past three years from NOAA to support his work on ENSO prediction. Of this amount, $38,000 was provided by the Office of Global Programs through a grant entitled "The development of probabilistic El Niño predictions." The balance was provided through the Office of Policy and Strategic Planning under a private consulting contract. Dr. Solow has received no support from the U.S. EPA in this area.

NOAA support to the Woods Hole Oceanographic Institution over the past three years is as follows:

```
1997 (through 6/97) - $1,169,942
1996                - $2,068,189
1995                - $2,333,449
```

Of this amount, only a small fraction supported research connected to ENSO. EPA support to the Woods Hole Oceanographic Institution is very small and not easily determined. In any case, none of it supported research connected to ENSO.

Chairman CALVERT. Thank you for your testimony.

Let me assure this panel that I think that long-term weather prediction is a responsibility of this government and certainly this Committee, and that we will be working toward assurances that the TOGA array will be funded this year with Chairman Rogers. And we expect that to happen hopefully very soon.

EL NIÑO AND WEATHER PREDICTION

Chairman CALVERT. With that, I would like to start off with some questions.

There is a big difference in California getting 150 percent of normal rainfall and 300 percent. How far would each of you go in making a prediction for this coming winter—knowing that it is what it is, just a prediction—using California, my home State, as an example?

Mr. BARNETT. Well, that's not a very nice question, frankly.

[Laughter.]

Mr. BARNETT. The answer is that at this stage our predictive capability has a lot of uncertainty. And I think the numbers you quote are probably the range of uncertainty or perhaps even more than that. So to get somebody to say it will be 220 percent or 110 percent 6 months from now or 3 months from now—I am sorry, there is not a good scientific answer to that question. I would like to give you one.

As I indicated, the probability during El Niño events that the southwest is going to be wetter than normal is around 70 percent. In a place like San Diego, which doesn't have a lot of rain, it is easy to get 200 percent of normal rainfall. In a place like San Francisco that does have a lot of rain, this becomes a much harder thing to do.

So I don't have a direct answer for you other than to say the probability of a wet season there—define it as you wish—is roughly on the order of 70 percent, again taking into account that what we're going to see here presumably is we only have one analog like it in the historical record. And you know what that did. So I would prepare for the worst, but be not awfully surprised if it didn't happen.

ADVANCEMENTS IN EL NIÑO PREDICTION

Chairman CALVERT. Carrying on with that for a moment, in the past predictions on El Niño have been more vague than they have been to this date, obviously. What has happened—what significant advancements have occurred that you become more specific than you have in previous years?

Mr. BARNETT. There are a number of things that contribute. As Mike Hall indicated, there has been a substantial ongoing research effort by his office. What it has resulted in are improved forecast models. We have the TAO array, which has been a tremendous help. And Ants Leetmaa, who heads the Climate Prediction Center, sitting behind me, I think would agree without the data from that array he would be hard-pressed to make a forecast that he had much confidence in.

So I think it is a combination of things: improved models, the TAO array, and scientific progress.

Chairman CALVERT. Yes, Doctor?

Mr. HALL. If I may add a thought to that reply, Mr. Chairman, one way of thinking about it is that a decade ago we were at the point of identifying the basic yes/no question: Was there one [Le Nino], was there not? Then we evolved into a state of being able to anticipate, to some extent, the basic yes/no question. There would be an event, or there would not. Later we came to realize that this isn't really about events; it is about a constantly evolving state between the cold phase in the Pacific and the warm phase. And we began to understand that the phasing of the event was fundamentally important. In other words, At what point were we in an event as it unfolds?

Finally, we began to realize and today can do something about the need to know the actual configuration of an event. What is the shape and intensity of the rainfall pattern in the tropics which affects global wind fields? How does the sea surface temperature in the tropics configure as a function of time? So that better models, better data to initialize the models, and a better understanding on the part of scientists of how these things unfold lead to better and better specificity in the prediction.

Even today the National Weather Service—and Dr. Leetmaa is prepared to come speak, if you wish—is developing specific information about how we imagine things unfolding in California in the coming several seasons. He will, of course, couch those estimates in probability terms. But he will suggest that the probabilities we are providing are considerably better than what decisionmakers would be using without these forecasts.

We expect now, as we learn about other types of variability, and learn to do a better job of tracking the shape—if you will, the configuration—of El Niño in the tropics for even the El Niño forecasts to improve.

OVERREACTION TO EL NIÑO

Chairman CALVERT. One other question before I pass it over to Mr. Roemer.

Is there any concern of overreaction? You said prepare for the worst. But sometimes we know the media sometimes can be very dramatic about certain stories. Time magazine had a story recently warning of landslides, flash floods, droughts, crop failures during this coming year.

Is this a problem? Do you think this is appropriate? Just your comments on that.

Mr. BARNETT. Worries the hell out of me, very frankly. I think what has been lost in the translation from the scientist and the people that do forecasting like myself to the press is the statement of uncertainty associated with it. Although it is given to the press, it oftentimes does not appear and people take it as a certainty that California will wash away this wintertime. I came back from Montana and my home was going to be gone, I was told. Come on.

I think there is an overreaction, yes. And I worry about that because for every swing to the left, there is one to the right.

Chairman CALVERT. Saying that, I move to my right to my friend, Mr. Roemer.

[Laughter.]

Mr. ROEMER. I see why you're Chairman with that segue. Very smooth, Mr. Chairman.

[Laughter.]

Mr. ROEMER. I thought the Chairman asked you a tough question, too, asking you to predict what the weather will be like for California. Certainly you probably can't predict that with any certainty, but does your predictive model tell us, for instance, who is going to win the World Series in baseball or anything like that?

[Laughter.]

Mr. ROEMER. We have a lot of tough questions like that.

Before I ask you some serious questions, I would like to yield to the distinguished lady from California, who I mentioned in my opening statement, Zoe Lofgren, for a comment.

Ms. LOFGREN. I appreciate that very much, Mr. Roemer.

I have another hearing going on at this same time with constituents from San Jose who are testifying, so I am going to be going back and forth. I did want to submit my statement for the record. I did want to indicate my support for funding this research activity, whether through a colloquy or—we will be working on a bipartisan basis to do that.

I think it is essential that we take advantage of the work you've already done and allow you to do additional work so that we are armed with information about our climate and take appropriate steps.

I understand that there is no certainty in the prediction. However, I am mindful—I live in northern California—they are catching blue marlin in the Monterey Bay. This is a very weird phenomena for us. The surfers are no longer wearing wet suits. We do understand because this has never happened in the memory of anyone I have talked to. So we know something odd is afoot and us Californians are repairing our roofs right now because we do suspect that it is a very strong possibility that something very dramatic could occur.

So I understand your concern about the press overplaying it, but really I think it would be wise for not just individuals but for government agencies to take appropriate steps to become prepared should that 70 percent chance occur. For example, the Army Corps of Engineers needs to gear up in terms of making sure that flood control channels can be cleared in a timely fashion instead of doing their regular paper shuffle. I mean, there are some things that we need to do.

I just wanted to say that I appreciate your letting me have a minute before I rush off and I will submit my statement for the record.

[The prepared statement of Ms. Lofgren follows:]

Statement of Congresswoman Zoe Lofgren

House of Representatives
Committee on Science
Subcommittee on Energy and Environment
Hearing on "Preparing for El Niño"
September 11, 1997

Thank you Mr. Chair, Mr. Ranking Member for yielding to me:

I think this hearing is well-timed. El Niño is back in California and around the world. I've heard anecdotal accounts of sport fishing for mahi mahi, tuna, and swordfish off the coast of Northern California. In the Monterey Bay, sea surfaces are reported to be between five and 10 degrees above normal.

Described as the "climatic event of the century", El Nino affects weather patterns around the globe. The current El Niño is compared to the last strong one in 1982-1983, which caused in California an estimated $265 million in damage and 14 deaths. In fact, the warming of the South Pacific is about four months ahead of temperature levels it had reached at this stage in 1982-1983. Experts predict that the California coast may face unusual weather this winter, including severe storms and warmer than normal temperatures.

I am very concerned about El Niño, and believe that we must be proactive to mitigate El Niño's economic and social consequences. That is why I'm so disturbed about the funding status of a NOAA effort to create an operational capability to predict and disseminate information on El Niño events.

This year the Science Committee authorized full funding for a forecasting system called TOGA (Tropical Ocean-Global Atmosphere program). It is my understanding that the House Appropriation Committee did not include funding to transition this program from a research enterprise to an operational program in the FY 1998 NOAA Appropriation. Given the recurring nature of El Niño, I believe that this funding denial is short-sighted.

When the appropriations bill comes to the floor, I intend to offer an amendment adding funds for this program. I hope I can work with colleagues on both sides of the aisle to rectify this oversight. I look forward with interest to the comments of today's witnesses.

EFFECT OF EL NIÑO ON MIDWEST

Mr. ROEMER. I thank the gentlelady, moving from blue marlins to the Midwest.

Dr. Hall, you mentioned in your testimony on page three and on page four, as we move away from California and we look at the rest of the United States, you talk about the flooding and the rainfall in Illinois. You say that flood damage was estimated at $100 million. On the next page, drought in the Midwest—again, I am from the great State of Indiana—you say that the corn crop yields dropped 50 percent from 1982 levels.

I would like you to be a little bit more specific on what kinds of consequences we might face in the Midwest—particularly in the agricultural areas—should El Niño sweep across the Midwest from California. What kinds of economic consequences do we face, particularly in the agricultural sectors?

Mr. HALL. Congressman, if I may respond and ask my colleague, Dr. Leetmaa, to say a word about the prediction in the Midwest, I will make a more general comment giving him a moment to think about an answer.

I want to point out that much of the evidence regarding climatological impacts in the Midwest is about summer conditions—the Mississippi floods of a few years ago, extreme droughts which can follow in the wake of strong El Niño events. The summer signal related to El Niño over North America is one of the most difficult things for us to get at. So the comments that we would provide you today will in fact be with even less certainty than you have heard about, for example, in precipitation fields over California in the winter season.

If I might, that points out very, very strongly the reason for looking at the other modes of variability that derive from tropical variations that force conditions over North American. Precipitation fields over North America are perhaps the one focused target of the research effort I have described that will unfold in the next 5 to 10 years.

Now having said that, I think the issue in the Midwest, and in the United States in general, in anticipating this thing that is coming in the winter, and in the following summer, is to try to build into our systems a resilience to the possible outcomes, not certain that those outcomes will happen, but trying to be resilient to their consequences, should they happen.

It is indeed a problem that climate prediction requires—absolutely requires—a relatively well-educated user. Any program of preparedness we mount has to have education of user as an integral component of the research program. Understand that what he has got is not certainty; it is merely a better probability than we had before we came along.

Having said that, if I might ask my colleague, Dr. Leetmaa, to describe to you what we expect as a consequence in the Midwest in the coming several months, we can come back to the detailed impacts of that if you wish.

Mr. LEETMAA. Mr. Chairman, may I?

Chairman CALVERT. Please.

Mr. LEETMAA. I am Dr. Ants Leetmaa. I am the Director of the Climate Prediction Center. This is the center that makes the official forecast. If the forecasts don't come true, you have me to blame.

For the upper Midwest, we anticipate above normal temperatures. For the wintertime, below normal temperatures. My understanding is that Dr. Solow in fact has looked at the agricultural impacts. So I am not qualified to talk about that. Perhaps you would like to.

AGRICULTURAL IMPACTS IN MIDWEST

Mr. SOLOW. I'm going to pass the buck a little bit, too. I think what the other speakers have said is true. The ENSO signal in the part of the country that you come from is not nearly as clear as it is in other parts of the country, like the southeastern United States, like California. So when we looked at the effects on agriculture of ENSO variability, the effect in that part of the country was relatively small compared to the effect in other parts of the country.

OVERSEAS EL NIÑO RESEARCH

Mr. ROEMER. Finally, Mr. Chairman, let me just conclude with one last question.

Do the Australians or the British—Are other countries ahead of us at all in terms of their predictive capability in this area? What are we learning from them?

Mr. HALL. I will make a quick comment and I suspect Dr. Barnett might want to comment in some detail.

First of all, we work really rather elaborately with modelling and prediction centers in other countries and the scientists that back them. A number of our efforts around the world at exploiting present information with respect to this event are in partnership with centers in other countries. There is going on this week in Zimbabwe a meeting of modelers from several centers around the world with affected governments in southern Africa to try to develop a consensus view—because after all the forecasts differ slightly—and provide to decisionmakers that consensus view.

The short answer to your question—I don't think anyone is ahead of us. I think there are a number that are at about the same point we are at and we are working very closely with them. The effort in the United States is strong. Indeed, I will go so far as to say it is the strongest effort in El Niño research and forecasting in the world. The United States has led the world in that regard and I hope we will continue to do so.

Mr. ROEMER. Dr. Barnett?

Mr. BARNETT. I generally agree with Mike's comment there, but some interesting things are happening in the world. The Europeans have formed a group to develop global climate forecast models. They view the world as one populated by a world economy that is sort of a cut-throat operation. Climate forecasts for them are strategic information that they can use, that their businesses can use.

You may also know that the Japanese have mounted a huge dollar effort, at least—they are just getting started—to take the lead in global climate forecasting. Already indications are that they know the economic value of those forecasts.

So I agree with Mike that we are well-positioned now, but we have some very good people that are going to be running hard right at us.

Mr. ROEMER. So all the more important that we make sure that we follow through on our bipartisan cooperation next week in assuring that this money, through the Appropriations Committee, is spent to help us be able to predict the effects of El Niño?

Mr. BARNETT. Yes, prediction. That's my thing. But as I said, what I am really here for is to say that we can do something now. We can truly do it better in the future. But we need to really learn how to use those. That is the hard thing to do. It is really a hard problem. Predict and utilize. Utilize.

Mr. ROEMER. Thank you.

Chairman CALVERT. Thank you, Mr. Roemer.

Mr. Ehlers?

VARIABILITY OF EL NIÑOS

Mr. EHLERS. Thank you, Mr. Chairman.

I would like to pose a question which all of you can answer individually and help educate me and the panel and get back to some of the fundamentals of what we know about the causes of El Niño. What is actually going on? I am interested in—for example, someone made the comment there is a current average of once every 4 years appears to be the average frequency, yet you say the last one was 1982. Are they really that irregular?

Secondly, the comment was made that 1982 was the only analog that we have to go by. Haven't we studied them thoroughly enough before that?

CAUSES OF EL NIÑO

Mr. EHLERS. But the broader question I am interested in is, what understanding do you have of the ultimate cause of the El Niño, first of all, and secondly, what is your understanding of how that affects climate and where does it affect climate? What is the extent of the change? Is it primarily just North, Central, and South America on the western half of the Nation? Is it worldwide due to the El Niño in the Pacific Ocean? What is going on?

VARIABILITY OF EL NIÑOS

Mr. BARNETT. To clarify a few things, the 1982–1983 event was the biggest in recorded history. And we are near that state now. So that was the reason for dredging that up. In fact, there have been El Niños in 1987, 1992, 1995—they are not periodic, unfortunately, or you could perfectly predict them. But they occur every 2 to roughly 7 years. Does that help with that?

Mr. EHLERS. So there is a big variability?

Mr. BARNETT. Yes.

Mr. EHLERS. It is 4 years plus or minus 2 or 3?

Mr. BARNETT. If you average over 20 or 30 El Niños, you will come out with an average of like 4 years, but there is variability.

CAUSES OF EL NIÑO

Mr. BARNETT. The other thing you asked was really an interesting question. What is the instant that starts an El Niño? What is the thing that does that?

I think frankly there is no agreement on that. Some think it is driven from the high latitudes of the southern ocean. Some think it is entirely a Pacific phenomenon. Some think there are signals that come from the Indian Ocean and trigger this. Some think signals come from the mid-latitude North Pacific. And probably when we really get down to it, they're all going to be right at one time or another. But the thing that just sets it off like that—I don't believe there is agreement on that.

GLOBAL IMPACTS OF EL NIÑO

Mr. BARNETT. Now where do El Niños impact? That is well known. Australia, eastern Australia, the wheat crop that was mentioned here, certainly South America, particularly northeast Brazil, South Africa, the West Coast of the United States for sure as well as the southeast part of the United States. What hasn't been dwelled on here—everybody has concentrated on El Niño, sort of the warm side of the coin. There is a cold side of the coin called La Nina also. In some ways, that is more devastating to the United States—particularly the farming interests—than is the warm event because you generally tend to have droughts in the summers over the central part of the United States when you have a La Nina and that is where your people can get in trouble.

Does that answer your question, sir?

REGIONAL IMPACTS

Mr. EHLERS. That answers part of it.

The climate effects—are they similar, for example, during El Niño—are they the same in North, South, and Central America? Or is there variability there that some have rain and some have drought? During La Nina, the same question.

Mr. HALL. I might take a beginning shot at that question, Congressman.

When an El Niño develops and unfolds, it changes the prevailing wind patterns over most of the earth. But it changes them in fairly complex ways. The result is that there are certain regions of the world where the effects of El Niño are fairly robust—what we would call robust—that is, you can count on them. The southeast portion of South America will be wetter than usual, in some cases very problematically wetter than usual. The northeast region of South America along the Atlantic coast will suffer drought. It is a very consistent signal. The southeast of the United States is pretty well regularly affected by El Niño in a way that we understand.

At the same time, the earlier question reveals in an instant—the question about the Midwest, summer versus winter, et cetera—reveals in an instant that there is a great deal about the global climate effects of El Niño that we don't know. Right now, we are beleaguered with questions about the drought in China, the drought that is affecting North Korea—how much of it is due to El Niño? All we can do is analyze the data, experiment with our models, and suggest that El Niño seems to have something to do with it. That is far from a capacity to predict it. That's a real-time analysis capacity, but it's much better than we had before.

These questions reveal to me how important it is to recognize that we know how to do certain things and there are certain things that we haven't yet figured out.

The world-wide effects of El Niño that we came to expect after looking at some 12 or 14 events in this century are unfolding in those robust regions pretty much the way we expected them to. On the other hand, you can tell that there are regions of the world where knowing that there is an event underway—and even knowing something about the tropical configuration of the event—its shape and its intensity—isn't sufficient to tell you much about what is going on. So the world is very uneven and the spacial distribution of our knowledge is very great.

Mr. EHLERS. At the very least we can be very thankful that's some oscillatory behavior rather than divergent behavior.

Thank you, Mr. Chairman.

Chairman CALVERT. Thank you, Mr. Ehlers.

Ms. Johnson?

NOAA OPERATIONAL CAPABILITY FOR EL NIÑO

Ms. JOHNSON. Thank you very much, Mr. Chairman, and thank you for having this hearing.

I am from the State of Texas that has every terrain and every climate and every weather condition. And certainly we have benefited from many of the predictions that have been made as it relates to weather. We in my area, which is north central Texas—Dallas, Texas—experience during the warmer climates probably more ozone alert days than any other place in the country.

And what I would like to ask you is, how would you reasonably manage the research? If you face a large amount of money being cut from your research activities, would you continue to a large extent or small extent or not at all? And do you have any sense of pay-back in an economic sense from establishing the operational capacity?

Mr. HALL. Let me try to answer that question.

I should remind the Committee of something it knows perfectly well, that 2 years ago in Fiscal Year 1995 the budget that sustains this and related climate work in NOAA suffered a rescission and thereby a reduction of $14 million, some 20 percent. At that time, that reduction had the effect of really riveting NOAA's attention on the things we cared about. It made us look very closely at the whole of our program. We chose to emphasize natural variability. We chose to emphasize mechanisms like El Niño—although not just El Niño—but mechanisms that affect climate fundamentally over a few years to a decade or two.

So we believe our research effort has been thoroughly scrubbed down, constantly reviewed, and is focused on the right problems. Therefore, the answer to your question in detail is this: In a reduced budget scenario, NOAA has to choose between emphasizing one of the two paths I described in my opening statement. In other words, emphasizing the operational capability we now have—which is useful but rather limited—versus developing an operational capability that really deals with patterns of variability over States like Texas. Texas has suffered severe drought recently. Texas may be looking at an exceptionally moist winter in the coming season. Clearly, this is a region that has a great deal to gain by our understanding this set of mechanisms far better.

The consequence of a reduction—failure to appropriate the money for the observational system would be an example of a reduction from a projected budget—will actually be fairly serious. We will have to diminish one of the two paths. I suspect NOAA will give substantial priority to the operational path, to pursuing the projections we now know how to do in an operational fashion. But it could mean literally many more years—if not a decade more—time in arriving at some insight about all of those mechanisms that affect the climate over Texas.

As for the reasons to do this—the economic justification—I think Dr. Solow's talk and others you will hear from on the next panel would suggest that we are already to the point where a cursory estimation of benefits would indicate that the pay-off is ten to one, and it is probably 100 to one or more. I heard someone use the figure earlier of 300 to one. I don't think that is a bad figure at all. Pretty good return on a research investment, pretty good return on an operational investment. That is why NOAA is so comfortable today urging that you take both paths, or that you enable us to take both paths.

Ms. JOHNSON. Did you wish to comment?

Mr. BARNETT. Only to add that the type of—I will call it an application program that I called for—need not trouble the national debt. Very frankly, a lot of the pieces of that are already in place and funded normally. What we really need here is an integration of people. So I think to begin to approach the kind of questions you ask is not a budget-breaking thing at all. It is a matter of the will to cross iso-empire lines and things like that and pull people together and figure out what information they really need and how they're going to use it.

Ms. JOHNSON. Thank you very much, Mr. Chairman.

Chairman CALVERT. Thank you.

Mr. Rohrabacher?

BENEFITS OF EL NIÑO

Mr. ROHRABACHER. Thank you very much, Mr. Chairman.

As a California surfer, I am very interested in El Niño and I will happily report to the Committee that my surfing friends are already experiencing El Niño and enjoying it and we would like it to stay around as long as possible.

[Laughter.]

Mr. ROHRABACHER. The water is warm and it is fun and actually fishermen are also telling me that they are catching a lot more fish out in California now.

So in terms of one segment of the population, El Niño is terrific. However, I understand there might be some other implications and some of the things you're talking about in terms of lack of water in other parts of the country and different climate repercussions in other parts of the world may not be as happy as the effects for California surfers.

HISTORY OF EL NIÑO

Mr. ROHRABACHER. First of all, I would like to ask whether or not El Niño is a phenomenon that is something that is relatively new. I understand that you have testified that perhaps we can recognize it has been around for 100 years. Has it been around for 1,000 years or 2,000 years?

Mr. BARNETT. Yes, certainly it is written about in the logs and memoirs of the Spaniards when they came into South America. People have looked in the glacial—looked at ice cores from Peru—yes, this is something that has been around quite a long time.

Mr. ROHRABACHER. Are we talking about thousands of years perhaps?

Mr. BARNETT. Almost likely, yes, but nobody knows for sure, of course.

Mr. ROHRABACHER. And El Niño is a totally natural phenomenon, then?

Mr. BARNETT. Yes.

EL NIÑO AND GLOBAL CLIMATE MODELS

Mr. ROHRABACHER. Has the impact of El Niño been included in the global climate models that are being produced in the last few years?

Mr. BARNETT. Only the most recent models have credible El Niños in them.

Mr. ROHRABACHER. So when people are trying to determine global warming, for example, they have not been adding the effects of El Niño into their models until recently?

Mr. BARNETT. No. The most recent models have credible El Niños in them. What I mean by that is that they produce an oscillation in the tropical water temperatures was about the right period, it has about the right spacial structure—but these are only the newest models.

Mr. ROHRABACHER. Right. I understand that some of the older models didn't even calculate the cloud cover in at certain times.

Mr. BARNETT. Oh, the cloud cover was constant. That field has come a tremendous distance in the last 10 years, incredible distance.

Mr. ROHRABACHER. Well, it should. We have spent enough money bringing it that distance.

Let me just note that I believe NOAA has done a terrific job in alerting the public and in studying this phenomena and is continuing to do a good job on this type of study of climate. And I would say that I fully endorse this type of study of the natural phenomena and how it affects humankind. I believe the work you folks are doing is valuable—I should say invaluable because it is absolutely—the information you are gathering will help us understand this environment that we live in and it helps us—not like today—but it will help us in the future.

As compared to global warming, which is something I would like to ask you about—global warming is based on the idea—global warming studies are based on the idea that climate change is taking place that is not a natural phenomena, that is based on human activity. And the study is basically being used as an excuse to regulate human activity.

El Niño, on the other hand—if I may be so bold—is actually a natural phenomena that we are studying to try to help us understand nature so that we can live better within our natural environment. Is that a good description of the difference between the two?

Mr. BARNETT. We have no control over it. We might as well learn to live with it.

FUNDING FOR EL NIÑO RESEARCH

Mr. ROHRABACHER. And let me just end my final question by—you might want to comment on this—do we have enough money being channeled into the study of these type of climate situations? Or is global warming draining off resources—is the study of global warming draining resources from the study of El Niño and other things that could be more worthwhile for us to put our money into?

Mr. BARNETT. Well, one thing that you need to realize is that the coupled climate models that we use to, (A) predict El Niño, and (B) predict its global impacts are the same models that people put CO_2 into to study the global warming problem. So they are not separable by any means.

Mr. ROHRABACHER. Yes, but you're sort of getting——

Mr. BARNETT. That is many for the price of one, if you wish.

BENEFITS OF EL NIÑO

Mr. ROHRABACHER. Okay. Well that is a fair answer.

Is there any other comment on that?

Mr. SOLOW. I wanted to—I think in the beginning of your remarks you raised an interesting point, and that is that El Niño, and ENSO generally, doesn't only cause damage, that it causes— it has climatic effects that may have benefits themselves.

Mr. ROHRABACHER. That's true.

Mr. SOLOW. And that part of the value of a prediction is that surfers or fishermen or farmers—if they are able to anticipate a solicitous climate—then they may be able to take measures in advance that will allow them to make the most of that opportunity. So it is important not to become preoccupied on the damages associated with these climate variations, but to also take into account the benefits of them directly.

Mr. ROHRABACHER. You can sell short on California wet suits for surfers or things like that.

[Laughter.]

Mr. ROHRABACHER. Your point is very well taken.

Thank you very much, Mr. Chairman.

Chairman CALVERT. Thank you.

Mr. Brown?

EL NIÑO AND GLOBAL WARMING

Mr. BROWN of California. I appreciate the thrust of the questioning by our good friend, Mr. Rohrabacher. And perhaps we could go just a little bit further with that.

Clearly, El Niño is a historical phenomenon that has been going on, as his questioning indicated, for quite a long period of time. Global warming—if in fact it does exist at all—is a more recent phenomena, what you might call anthropormorphic, generated by human beings. Mr. Rohrabacher has been considerably more skeptical about that. And he may well be right, but I think we need evidence to demonstrate whether he is right or not right in this particular situation. The question that I would have is kind of the reverse of what he was asking, if I understood him correctly.

If the more recent phenomenon of global warming truly exists, is there any way of detecting whether it might have an impact on either the frequency or the—shall we say level—of the El Niño? In other words, could there be a feedback relationship between these two which either might amplify or reduce one or the other? Is that the kind of question that could reasonably be asked? Could the research reasonably provide some inklings of an answer to that question?

Mr. BARNETT. Yes, I think that's mine here, an excellent question. Just an excellent question.

The best results to date that I am aware of show that as you gradually increase the CO_2 concentrations of weather pollutants in the models we have been discussing, the models that produce an El Niño, you don't see any significant change in the behavior of the El Niño. Remember, these are studies made with computer models. Unfortunately, that is the only way we have to study this phenomena. They are the best we have today, although imperfect, but their message is relatively clear on this score.

Mr. BROWN of California. I'm looking at the charts generated by one of these computer models. It is labelled "Multi-varied ENSO Index for the Six Strongest Historic El Niño Events Versus Current Event." While the current event is just beginning, it is clear from the chart that it is substantially above any of the previous events that appear here.

Now to an uninformed layman, this could indicate that if there really is an impact or relationship between El Niño and global warming and that global warming is increasing, we can automatically jump to the conclusion that the current El Niño shows this greater impact because it is being reinforced by global warming.

Are you saying that there isn't enough evidence to justify that?

Mr. BARNETT. Certainly not. No scientist would take the sample of one. But I also might mention—I think it was 1918—somebody can correct me on the year [1877/78]—there was a mammoth El Niño comparable to the two we're talking about now. Certainly nobody would argue then that global warming was responsible for that.

Mr. BROWN of California. No. That's why I am asking for reliable scientific advice to weigh these——

Mr. BARNETT. I'm not sure you got it, but at least that is my opinion.

[Laughter.]

Mr. BROWN of California. I don't want to interpret Mr. Rohrabacher's position, but I think the view he is taking—a perfectly legitimate one—is that we have jumped the gun on the impacts of global warming and are preparing to take certain economic steps that might have a major impact without the major supporting evidence that there really is this kind of phenomenon. Am I reasonably correct in this, Mr. Rohrabacher?

And that he probably doesn't object—in fact, he seems to point with pride at the research itself having to do with both El Niño and to a lesser degree with global warming. If we could just find out that global warming impacted surfers, he might enthusiastically adopt that.

[Laughter.]

Mr. BARNETT. Sir, this is a complicated issue, I think, that goes beyond the scope here. I will say that I was co-author on the last two Intergovernmental Panel on Climate Change (IPCC) chapters on detection of anthropogenic signal. It is a complicated issue and one that I don't think we should maybe—if you want to take time, we can do that. But there is work going on there that is very exciting. Hopefully in the next year or so we can dispel or reinforce some of the uncertainties that you have. It is quickly leaving the realm of scientific discourse, as you are well aware, and all poor people like me can do is give you our best guess based on science.

ANTHROPOGENIC SIGNAL

Mr. BROWN of California. Well, I value the contribution Mr. Rohrabacher makes to this argument because he may well be right and we just do not yet have enough information to conclude that. But if you are correct, we will have more such information in the relatively near future. Is that what you're saying?

Mr. BARNETT. I was one of the first guys that did a hard-core detection study back in the 1980's and asked the question, can you see the predicted [greenhouse] signal in the data? The answer was a slam dunk "no." That study was repeated with improved techniques by other colleagues in the early 1990's and the answer wasn't quite as clear a slam dunk, but "probably not." The study has been continued on now with even better techniques, better models, better data. The answer is, "uh-oh, there might be something there." We are sort of just coming out of the noise, if you wish.

So it is argumentative certainly from a scientific point of view, but I think that is the history of it in a nutshell.

Mr. BROWN of California. Well, I hope we can pursue this in further discussions and hearings. I have a great deal of respect for Mr. Rohrabacher's views. I think he has some respect for mine also. Maybe we can reach a meeting of the minds here.

I think the question of whether we should intervene to reduce— if it's possible—the impact of global warming is an open question and probably can't be answered on the grounds of what the actual impacts of global climate are going to be. But as I understand it, many or most of the proposals being made—for example, the reduction of fossil fuel burning—have benefits independent of their impact upon global warming, per se, and can be justified on those grounds.

I yield back the balance of my time, Mr. Chairman.

Chairman CALVERT. I thank the gentleman. I am certain that as time goes on we will have future hearings on the subject of global warming.

Mr. BROWN of California. I am sure we will.

Chairman CALVERT. To the interest of all.

With that, Mr. Roemer, I believe you have an additional question.

BENEFITS OF EL NIÑO

Mr. ROEMER. I have one final question that has just been sparked by Mr. Rohrabacher and then rereading parts of your testimony, Dr. Solow.

You talk about—and Mr. Rohrabacher talked about the possibility of some benefits—economic benefits or surfing benefits to a certain segment of the population here. You talk a little bit more in terms of regional or sectoral benefits, that there might be some.

Apart from what it might do for warmer water and waves for surfers, both in El Niño and La Nina, are there some benefits that we might see to the East Coast, for instance, in lower heating bills when El Niño comes in? Can you be more specific on El Niño and La Nina in terms of regional benefits?

Mr. SOLOW. One substantial example, as I mentioned in my testimony, is that an El Niño event is associated with reduced tropical storm activity in the Atlantic. The number of hurricanes this year is very low. It is very quiet out there. That is associated with El Niño. So there is an example of how El Niño could do something that we like.

I would just say generally that all variations in climate aren't bad. If you have a drought under one phase of ENSO and you have more rainfall under the other, one may cause damage to agriculture and the other will cause benefits to agriculture. I believe our study showed that during cold events agricultural production was higher, particularly in the southeastern United States. So that is a situation in which farmers, if they are prepared for that, can take advantage of that and bring home a good crop, while under an El Niño event, they may not be able to.

USING EL NIÑO PREDICTIONS

Mr. ROEMER. But it all comes back to Dr. Barnett's comment, too. And I will conclude here so we get to the next panel.

Dr. Barnett emphasized that we not only have the ability to predict, but we have the ability to utilize. Certainly many of the good benefits cannot be utilized unless you get that information out to people and they are aware of what happens potentially on the good sides. We have concentrated 90 percent in this hearing on the bad sides of El Niño. But certainly we get back to the question of prediction and utilization.

Mr. SOLOW. I would just say that a lot of this economic activity is carried out by fairly large and sophisticated firms. That includes agriculture but also power generation. And firms in those industries are keenly aware of the effects of weather and climate variation and are ready to begin to use these predictions. So they are not totally in the dark about this.

Mr. ROEMER. Very interesting.

Thank you again for your testimony.

Chairman CALVERT. I want to conclude this panel by saying thank you very much. Your testimony was extremely interesting. We look forward to this coming season with great interest to see how those predictions come true or not true, depending on what occurs. But certainly we thank you for your travels. And I know it's tough coming from La Jolla, California. You probably have the most beautiful office in the world, Dr. Barnett. I have been there, so thank you.

PANEL II

Chairman CALVERT. Our second panel will address the reaction of federal and local agencies to the El Niño prediction. As they take their chairs, I will talk about them.

Michael Armstrong is the Associate Director of the Office of Mitigation for the Federal Emergency Management Agency, better known as FEMA; Dr. I. Miley Gonzalez is the Under Secretary for Research, Education, and Economics at the U.S. Department of Agriculture; and Douglas Wheeler as Secretary for the California State Resources Agency.

Gentlemen, if you would please rise to be sworn in. Raise your right hand.

Do you swear to tell the truth, the whole truth, and nothing but the truth, so help you, God?

[All witnesses respond affirmatively.]

Chairman CALVERT. Thank you very much. You may take your seats.

Again, as with the first panel, without objection your full written testimony will be printed in the record, but I would ask that you summarize your opening remarks in 5 minutes.

Mr. Armstrong, you may begin.

TESTIMONY OF MICHAEL ARMSTRONG, ASSOCIATE DIRECTOR, OFFICE OF MITIGATION, FEDERAL EMERGENCY MANAGEMENT AGENCY, WASHINGTON, DC

Mr. ARMSTRONG. Good morning.

Mr. Chairman and distinguished members of the Subcommittee, my name is Michael J. Armstrong. I am the Associate Director for Mitigation with the Federal Emergency Management Agency. On behalf of FEMA and its director, James Lee Witt, I want to thank you, Mr. Chairman and the members of the Subcommittee, for inviting FEMA to be here today to discuss preparations for El Niño.

In my written testimony you find that we have enumerated some of the programs at FEMA that we think can and have been mobilized to respond to concerns about El Niño. I would like to emphasize several things in particular.

Firstly, that our emergency management system is built on a partnership of local, state, and Federal Government as well as voluntary agencies, businesses, industry, and individual citizens. All these folks are focusing on saving lives and property and reducing the effect of disasters.

At the federal level we are working very closely with NOAA. Based upon NOAA's forecast, there is ample information to indicate that the effects of El Niño may present us with extreme weather conditions, posing increased risks of disasters due to flooding, mud slides, and possibly hurricanes. NOAA's forecasts for El Niño are being used by the Federal Insurance Administration—which is part of FEMA—in its marketing efforts to promote the sale of flood insurance, especially in the California area. The information and resources pertaining to NOAA's El Niño research activities are readily available on the Internet and FEMA takes those predictions seriously and is grateful for their analysis and their predictions.

In my written testimony we talk about the different programs at FEMA that we think have a direct relationship to the El Niño concerns. My area of FEMA is mitigation, which is the risk reduction and prevention part of FEMA that works in both pre-and post-disaster scenarios to try to get buildings and structures either constructed in safer areas or elevated or relocated out of harm's way. In particular, the mitigation programs that could have an impact are post-disaster programs, including our hazard mitigation grant program, as well as our pre-disaster programs, such as the flood mitigation assistance program, and the new disaster-resistant communities initiative.

Our response and recovery directorate works specifically with state and local first responders and state offices of emergency management to prepare for events should they occur and support the initial response and recovery efforts in coordinating other federal agencies under the federal response plan.

And finally, our preparedness training and exercises directorate works through its Emergency Management Institute in Emmitsburg, Maryland, with training facilities and training programs for state and local emergency managers and other related disciplines along with hazardous material programs, beefing up state and local planning efforts, and working with general preparedness issues.

With the permission of the Chairman, I would like to supplement my written testimony—amend that testimony with a memorandum from our acting regional director in our region nine office in San Francisco. And I would like to highlight some of the things that are specifically going on in that region. Region nine covers the States of California, Nevada, Arizona, Hawaii, and the Pacific territories. Region nine has already approached the States of Nevada and California and it will be meeting with Arizona as well to coordinate and plan for the possibility of simultaneous incidents in all of those areas based on what we learned from the 1982 El Niño. Both California and Nevada have asked for training in several specific program areas, including preliminary damage assessments, program eligibility, and damage survey report preparation.

The regional response and recovery office has agreed with California to have a working document within 60 days that outlines procedures to standardize response operations. A parallel initiative is being implemented in Nevada and Arizona and will be extended to support Hawaii and the Pacific areas. And the various divisions in the FEMA regional office continue to be actively involved in the Governor of California's flood emergency action team.

One of the things that is occurring out there in Region nine is that the staff at the disaster field office in Sacramento—which is still actively pursuing the response and recovery efforts from the last series of flood disasters—are working on reducing processing time for typical small disaster survey reports to 48 hours or less.

Our disaster close-out center—there was a meeting last month, August 22nd, with the California Governor's Office of Emergency Services. That office agreed to supply a list of highest priority flood-related projects.

We also have regional environmental officers that have now been on-staff for the last year or so. Among the projects the environmental officers are pursuing is to develop an expedited review process to clear and close outstanding public assistance and hazard mitigation grant program projects in both Arizona and California. We think that if we can expedite those programs that those programs will in fact support efforts that may be of a concern to El Niño.

Also the Preparedness Training and Exercises Division has incorporated the California Response Information Management System, RIMS, and the application into the regional operations center in support of state standardized emergency management system. The FEMA region just finished hosting the 8th annual "Continuing the Challenge Hazardous Materials Emergency Response Workshop" in Sacramento with about 1,000 attendees. One of the issues addressed there was responding to El Niño.

Our public affairs operation is working on an El Niño loss reduction center on the FEMA web site, emphasizing preparedness and education. We are coordinating a Sacramento/Delta/Los Angeles River zone education campaign with El Niño efforts with multilingual and multicultural approaches. And we are working with the Federal Insurance Administration to place radio advertisements in the Los Angeles, San Francisco, and San Diego markets. Radio scripts in English and Spanish will reflect the El Niño prediction and encourage the public to purchase flood insurance. The campaign will begin September 15th and run for 3 weeks. And more placements will be scheduled in the next fiscal year.

We are pursuing other efforts as well that will be part of the record that we are amending for you.

[The prepared statement of Mr. Armstrong follows:]

Statement of

Michael J. Armstrong

Associate Director for Mitigation

Federal Emergency Management Agency

Before the

House Subcommittee on Energy and Environment of the
Committee on Science

September 11, 1997

Introduction

Mr. Chairman and distinguished members of the Subcommittee, my name is
Michael J. Armstrong, and I am the Associate Director for Mitigation with the Federal
Emergency Management Agency. On behalf of the Federal Emergency Management
Agency and its Director, James Lee Witt, I want to thank you, Mr. Chairman and
members of the Subcommittee for inviting FEMA to be here today to discuss
preparations for El Niño.

While we are best known for our response to disasters the opportunities presented by the
Nation's attention on the El Niño point to some of our other critical responsibilities. The
Nation's emergency management system is built on a partnership of local, state and
federal governments, voluntary agencies, businesses, industry and individual citizens
focused on saving lives and property and reducing the effects of disasters. Activities
surrounding El Niño afford us the opportunity to make that partnership.

At the federal level we are working closely with NOAA. Based upon NOAA forecasts
there is ample information to indicate that the effects of El Niño may present us with
extreme weather conditions posing increased risks of natural disasters due to flooding,
mudslides and possibly hurricanes. NOAA's forecasts for El Niño are being used by the
Federal Insurance Administration in its marketing efforts to promote the sale of flood
insurance especially in the California area. The information and resources pertaining to
NOAA's El Niño research activities are readily available on the Internet. FEMA takes
these predictions seriously and is grateful for the analysis and projections provided by
NOAA.

NOAA is an important FEMA partner providing invaluable contributions in their role
with assessment, prediction and warning activities associated with extreme natural hazard
events. These actions further public safety, preparedness and mitigation for the nation.
FEMA and NOAA share several successes in establishing interagency coordination
within programs in states where FEMA programs impact coastal areas, where more than
50% of our Nation's population resides and where an increasing concentration of our
population will be moving in the future.

Other examples of current cooperative activities include FEMA involvement with NOAA
in U.S. efforts for the Year of the Ocean. NOAA has provided advice and technical
assistance for FEMA's ongoing evaluations of erosion hazards study mandated by the
National Flood Insurance Reform Act of 1994. NOAA is also included as an advisor in
several other ongoing FEMA activities such as the Technical Mapping Advisory Council
and the National Flood Insurance Program Community Rating system. Together, NOAA
and FEMA are conducting a series of regional hazard mitigation conferences to improve
coordination among state program offices supported by federal programs. These
programs include state coastal zone management programs authorized under the Coastal

Zone Management Act, the National Flood Insurance Program and hazard mitigation activities authorized under the Robert T. Stafford Disaster Relief and Emergency Assistance Act.

Under FEMA Director James L. Witt's leadership, FEMA is completely changing the way the Agency does business. This is being accomplished by ensuring that communities and states put in place effective programs of mitigation and preparedness in order to reduce disaster costs and improve response and recovery operations. If a community has implemented sound mitigation practices in order to reduce the effects of disasters, as well as effective preparedness programs to ensure that emergency responders are trained and exercised in critical response and recovery operations, then the community is far better protected against the threats it faces.

Many FEMA programs directly support state and local efforts to prepare for El Niño. What follows highlights these efforts.

Mitigation

In recent years there has been an increase in major disasters including Hurricane Andrew, the Northridge Earthquake, the Midwest Floods of 1993, Hurricane Fran and others. Over the past five years, the average annual cost to FEMA alone has been $2.4 billion. As the occurrence of major natural disasters increases it has become evident that the solution to reducing future disaster costs is mitigation. Natural hazard mitigation efforts throughout the nation have been increasing significantly in recent years, with leadership being provided at the federal level. The passage of amendments to the Stafford Relief Act in 1993 which increased the level of funding available for mitigation after a disaster, was a turning point in the Nation's approach to disaster recovery. Instead of putting people back into harm's way - the way they were prior to the disaster - FEMA has emphasized mitigation and preparedness as cornerstones of emergency management.

The significance of El Niño for citizens living in harm's way is that once again they are reminded of the risks they face and are alerted to their options, yet many are still unable to or choose not to take precautions. Through various avenues described herein, FEMA is responding to this condition by mobilizing a very aggressive mitigation implementation effort. For those individuals and communities ready to take action, FEMA has a variety of mitigation opportunities from which to choose.

Hazard mitigation is defined as any sustained action taken to reduce or eliminate the long-term risk to human life and property. FEMA has developed a National Mitigation Strategy which is intended to strengthen partnerships among all levels of government and the private sector to ensure safer communities.

Pre-Disaster Mitigation Opportunities

One of the most important pre-disaster steps a state or community can undertake is the development of a comprehensive state hazard mitigation plan. The development of these plans results in states and communities having to identify their hazards, investigate various mitigation options and set forth priorities. These plans are critical to a state establishing a strategy to contend with disasters and FEMA provides ongoing assistance with the development of these plans.

Flooding, in various forms, poses the greatest threat by El Niño. Currently approximately 19,000 communities nationwide participate in the National Flood Insurance Program. As participants these communities have adopted flood hazard maps which are the basis of local floodplain management planning. Participating communities review new development proposals to assure that design and construction techniques are protected from the flood hazard. Since the beginning of 1975, over 2 million buildings have been built in special flood hazard areas incorporating a protection level tc the 100-year flood. FEMA estimates that each year the community floodplain management ordinances prevent over $770 million of flood damage to buildings and their contents.

The purchase of flood insurance is largely considered the most immediate step individuals can take to protect themselves from the threat of increased flooding associated with El Niño. A sound protection plan for a building should include both flood insurance coverage to protect against financial losses and a package of mitigation retrofitting options for the structure to reduce the risk of damage. The Federal Insurance Administration has a specific radio marketing plan scheduled to begin September 15 that reflects the predictions made based upon El Niño and the ad encourages people to buy flood insurance. The script will be in both English and Spanish and will target the Los Angeles, San Francisco and San Diego areas. A flood insurance claim provides a greater benefit than a disaster assistance loan or emergency grant. Anyone can buy flood insurance as long as their community participates in the program.

With the adoption of the National Flood Insurance Reform Act of 1994, Congress provided a new mitigation tool, The "Flood Mitigation Assistance Program." FEMA's Flood Mitigation Assistance provides funding to assist states and communities in implementing measures to reduce or eliminate the long-term risk of flood damage to buildings, manufactured homes, and other structures insurable under the National Flood Insurance Program. The Flood Mitigation Assistance provides planning grants to assist with updating Flood Mitigation Plans and Project Grants to implement measures to reduce flood.

Post-Disaster Mitigation Opportunities

The Hazard Mitigation Grant Program (HMGP) under Section 404 of the Robert T. Stafford Disaster Relief and Emergency Assistance Act represents the most prominent post-disaster mitigation funding source. Since the increase in funding available through this program as a result of amendments to the Stafford Act following the Midwest Floods of 1993, numerous communities nationwide have carried out mitigation activities which result in safer communities. For example, following the Midwest floods of 1993, approximately 5,100 repetitively flooded-properties were acquired or elevated, in the 30 highest repetitive-loss communities of Illinois and Missouri, with an estimated project cost to FEMA of $66.3 million. The cost of acquiring or elevating these properties was approximately 35 percent of total flood insurance claims paid over the previous 17 year period studied.

FEMA continues to strive to improve the grant program. Based upon several years of experience and valuable input from our state partners FEMA has embarked on an aggressive plan to streamline the HMGP and expedite processing of grants. Highlights of this effort include the following: expansion of the eligibility of funds for future disasters throughout the state when a presidential declaration has been made; delegation of National Environmental Policy Act review authority to FEMA Regional Offices and states, which will expedite review; simplification of the appeals process; establishment of closeout deadlines which will be supported by technical assistance teams; development of improved guidance materials to include a desk reference, applicant's handbook, training materials and job aids; deployment of benefit-cost training sessions for FEMA Regional Offices and States; and, preparation of a more advanced training curriculum in cooperation with State Hazard Mitigation Officers.

Additionally, through the National Flood Insurance Reform Act of 1994, new policy coverage has become available. The "Increased Cost of Compliance" coverage provides for the payment of a claim to help pay for the cost to comply with State or community floodplain management laws or ordinances after a flood event in which the building has been declared substantially damaged or repetitively damaged. The new coverage will help pay for the cost to elevate, floodproof, demolish, or relocate the building up to a maximum benefit of $15,000. This new coverage has just become effective as of June and we are very excited to see the role this will play as a mitigation resource for policy holders.

Mitigation through Partnership Alliances

Establishing successful cooperative partnerships with local, state and private sector partners is the key to FEMA's approach in implementing successful mitigation opportunities. To this end FEMA has taken several steps to involve these partners in preparing for mitigation implementation.

FEMA Regional Offices conduct Mitigation Partnership meetings with representatives of state emergency services and floodplain management agencies. These partnership meetings are all-hazards in their orientation and involve pre- and post-disaster mitigation opportunities.

At the national level FEMA seeks out and depends upon the input of representatives from professional associations on matters related to mitigation policy development and program delivery. Organizations such as the National Emergency Management Association and the Association of State Floodplain Managers represent local and state mitigation practitioners who have invaluable contributions to offer in the interest of making mitigation workable and realistic at the local level.

In the private and nonprofit sectors FEMA is working hard to build mitigation alliances with the Red Cross, the Institute for Business and Home Safety (formerly the Insurance Institute for Property Loss Reduction), the National Homebuilders Association, the American Society of Civil Engineers and other groups who influence or have a role in how we approach use of high hazard areas.

Mitigation through Disaster Resistant Communities

Recognizing that the only way to reduce the effects of natural hazard events such as El Niño is through mitigation, FEMA's new Disaster Resistant Communities initiative was developed. It is based upon a foundation of extraordinary partnerships and the use of incentives. In 1997, FEMA received $2 million from Congress to support pre-disaster mitigation programs. The Administration has requested $50 million from Congress for fiscal year '98 to continue the Disaster Resistant Communities initiative. This request is currently pending in conference. The initiative is designed to move communities from being victims of disasters to taking control of their future through effective pre-disaster mitigation activities. The Disaster Resistant Communities initiative has identified seven communities to act as pilots for this process. For each community selected an agreement will be signed by FEMA, representatives of the private sector and other stakeholders that commits all signatories to provide resources and support necessary for that community to become more disaster resistant. These pilot projects will illustrate the cost effectiveness of mitigation and show the partnership process needed to implement mitigation measures at the local level.

Preparedness and Training Activities

A joint meeting was recently conducted between NOAA and FEMA in order that FEMA could be fully apprised of NOAA's forecasts and predictions. Based upon these forecasts

FEMA is preparing to meet with state emergency management agencies representing states from California to Florida along the southern tier of the country. The purpose of meeting will be is to review preparedness issues and determine readiness activities that may be appropriate.

Although the impacts of El Niño may be widespread across the nation, the forecasts appear to focus on California as the most likely site for extreme rainfall events. Recognizing this, the following information is provided to reflect various preparations being made between FEMA and the State of California. Because of the wide range of threats that California faces on a regular basis, the State and local governments have, over the years, worked very closely with FEMA to build a very effective emergency management infrastructure that is integrated at all levels. FEMA works very closely with the State of California in the development of this capability. This involves regular meetings with State emergency managers to provide technical advice and assistance, determine functional areas that require additional emphasis throughout the year, training, and the conduct and evaluation of exercises.

With the beginning of each Fiscal Year, FEMA and the State of California negotiate and sign a Performance Partnership Agreement (PPA) which outlines those areas of mitigation, preparedness, response or recovery that need particular attention and lays out the parameters of the working relationship. In addition, the State and FEMA also sign a previously negotiated Cooperative Agreement that provides Federal funding in support of these functions. For example, in Fiscal Year 1997, FEMA awarded grants to California totaling $10,301,973 under the annual Cooperative Agreement to support mitigation, preparedness, and response and recovery activities. These funds are used for a variety of purposes, including the revision of emergency plans and procedures to better implement California's Standardized Emergency Management System (SEMS), the delivery of training, the conduct of exercises and upgrades to radio systems.

FEMA recently introduced the Capability Assessment for Readiness (CAR) process, which allows FEMA and the respective States to work together in identifying both effective emergency management processes and deficiencies that need to be corrected. FEMA has completed the CAR process with the State of California (as well as Oregon, Washington, Nevada, and Arizona), and has been working with both California and the other West Coast States to ensure that necessary corrective actions are taken to improve response capability in the event of a disaster. California has taken the initiative, along with other States, to expand the CAR process to include local governments. The results of both the State and local survey will significantly enhance response capability by eliminating deficiencies that could have slowed response operations.

California actively participates in FEMA's emergency management training program. During Fiscal Year 1997, for example, 560 students from California were enrolled at FEMA's Emergency Management Institute in Emmitsburg, Maryland. In direct preparation for El Niño, the California State Training Institute will have trained approximately 150 people in the Hazardous Weather and Flooding course, with sessions being offered for coastal, southern and northern California audiences. In addition, two sessions of the Flood Fight Operations and Expedient Flood Training courses were scheduled.

In order to assist emergency managers across the country, FEMA's Emergency Management Institute and the National Weather Service have formed a partnership to develop and conduct training. We have deployed:

- The Hazardous Weather and Flood Preparedness Course, which is a 2 ½-day field course that teaches basic hydrology and meteorology, what forecast information is available, how it can be obtained, and how it can be used in conjunction with local conditions to protect potential impacts.

- Partnerships for Creating and Maintaining Spotter Groups Workshop, which is designed to foster joint National Weather Service and emergency management spotter group planning, training, and operations for enhanced community warning for developing weather threats.

These courses are available for offering in California and other States on the West Coast that could experience the effects of El Niño.

Exercises are a critical part of our efforts with California and other States throughout the country. Exercises are designed to ensure that personnel, plans, procedures, communications, and emergency operations facilities have been tested in simulated conditions to verify that they are working effectively and are ready for emergency activation and implementation. California has an intensive State and local exercise program that requires the regular conduct of exercises as well as correcting identified deficiencies to improve State and local readiness.

FEMA's Office of Emergency Information and Media Affairs is taking several steps to promote preparedness and mitigation opportunities associated with El Niño. These include: establishing a new element of our website entitled "El Niño Loss Reduction Center" with links to material from the Federal Insurance Administration, Mitigation, Preparedness and the National Weather Service; preparing El Niño information packets for Congressional use including content suitable for a member newsletter or press release; meeting with NOAA's lead Public Affairs staff person responsible for El Niño issues; and, entered into a partnership with a major internet provider to further disseminate mitigation and preparedness information.

Response and Recovery

Should a major disaster--or multiple disasters--occur as a result of the effects of El Niño, FEMA and our Federal partners stand ready to respond. Since 1992, FEMA has utilized the *Federal Response Plan* to guide the response operations of twenty-seven (27) Federal departments and agencies and the American Red Cross. Each disaster has presented substantially different challenges, and each was utilized as a learning experience for the Federal government. Disaster victims have diverse needs, as do the State and local governments that are attempting to recover from disasters.

Regarding the potential effects of El Niño, FEMA Region IX has already contacted the States of California and Nevada to begin planning for the response to and recovery from a significant flooding event, based upon the effects of the 1982 El Niño. Both States have asked for specific training on Preliminary Damage Assessments, program eligibility, and Damage Survey Report preparation. In cooperation with each of these States, Region IX is preparing a schedule for initial training sessions, which will include both State and local personnel. Similar meetings and training sessions will be conducted with the State of Arizona.

In addition, Region IX is working cooperatively with the State of California in an attempt to prioritize any appeals, supplemental requests, and Damage Survey Reports that pertain to prior flood related damages, and to process these requests as quickly as possible. The California Office of Emergency Services will be providing FEMA with a list of their highest priority flood-related projects, and FEMA will make every effort to expedite these projects.

FEMA is constantly updating and streamlining its programs, procedures, and regulations in order to meet the needs of the next disaster. For example, FEMA has fully automated our Individual Assistance Grant programs, utilizing a 1-800 phone number to accept applications from disaster victims. The primary benefit of the computerized registration form is the speed with which assistance is now delivered to the disaster victims. Transfer of the registration to the processing center now occurs within hours, and in some cases just minutes, from the time the initial registration is taken. In many disasters, eligible victims will receive Federal assistance within 10-15 days after submitting their initial application.

Over the past four years, FEMA has been no stranger to large, multi-state disasters. In 1993, nine (9) States were severely affected by the Midwest Floods. As a result of Hurricanes Erin, Opal, and Marilyn in 1995, there were six (6) major disasters declared in the states of Alabama, Florida, Georgia and the Virgin Islands and the Commonwealth of Puerto Rico. In 1996, Hurricane Fran severely impacted the State of North Carolina, as well as impacting the states of South Carolina, Virginia, West Virginia, Maryland, and Pennsylvania. Most recently, flooding in the upper Midwest along the Red River severely impacted the States of North Dakota, South Dakota, and Minnesota. In each instance, the Federal government, utilizing the *Federal Response Plan*, was able to effectively respond to the needs of both disaster victims and the affected State and local governments.

Conclusion

In closing, I would like to reiterate FEMA's readiness to respond to emergency needs brought on by El Niño, and do want to reaffirm Director Witt's view and leadership in asserting that it is only through an aggressive nationwide movement to embrace and seek out mitigation opportunities, supported by effective preparedness programs that future disaster costs will be significantly reduced.

Again, thank you for this opportunity to share these views.

**Federal Emergency
Management Agency**

Michael J. Armstrong: Associate Director for Mitigation

On April 28, 1997, Michael J. Armstrong was nominated by
President Clinton to be Associate Director for Mitigation at
the Federal Emergency Management Agency. The nomination
was confirmed by Congress on June 12, 1997. Mr. Armstrong
was Regional Director of FEMA's Region 8 since January 1994.
As Regional Director, he coordinated mitigation, preparedness and
disaster response and recovery activities in Colorado, Montana,
North Dakota, South Dakota, Utah and Wyoming. His region was
cited as a "Center of Excellence" in developing national policy for
community relations and outreach. Mr. Armstrong successfully
initiated partnerships with the private, non-profit and public sectors
to break the recurring disaster-recovery-disaster cycle.

Before joining FEMA, Mr. Armstrong served for more than a
decade in local and state government. His previous position was
as Deputy Director of the Governor's Office of Energy Conservation.
Earlier in his career, he worked for 10 years as an assistant attorney
in Aurora, Colorado, where he served as a prosecutor and as a specialist
in personnel matters and land use issues. Early in his professional career,
Mr. Armstrong worked as a newspaper reporter.

A resident of Arvada, Colorado, Mr. Armstrong holds a B.A. degree
in English and a B.S. degree in journalism from the University of
Colorado at Boulder, and a J.D. from the Pepperdine University of
Law in Malibu, California. He serves on several non-profit boards.

Michael J. Armstrong

Chairman CALVERT. Thank you.

Dr. Gonzalez?

TESTIMONY OF I. MILEY GONZALEZ, UNDER SECRETARY FOR RESEARCH, EDUCATION, AND ECONOMICS, U.S. DEPARTMENT OF AGRICULTURE, WASHINGTON, DC

Mr. GONZALEZ. Thank you, Mr. Chairman.

I am very pleased to be invited to participate in this important hearing because fluctuations in weather and climate are major causes of variation in crop yield and commodity prices. In the interest of time, I will summarize my testimony and have submitted a full statement for the record.

Before I begin, let me introduce Mr. Al Peterlin, our chief meteorologist at USDA. I may call on him to help me answer some of the questions since I am not a scientist in this particular area.

Mr. Chairman and members of the Subcommittee, I know you agree with me that farmers and ranchers are hard workers and good managers. Food and fiber production, however, is a high-risk enterprise. The uncertainty of an El Niño phenomenon, which disrupts global rainfall and wind patterns can make it even more so by playing havoc on agricultural production. Research, economic analysis, and transmittal of new knowledge about weather and climate variability, therefore, are critical to us in our day-to-day activities and how we serve our customers.

It is now fairly well recognized in scientific circles that the interaction of atmosphere and ocean, primarily in the tropical Pacific, has a profound influence on global weather. There is ample evidence in published literature that El Niño-Southern Oscillation, known as ENSO, can influence significant shifts in regional climate over some regions of the continental United States, especially along the West Coast and Gulf Coast regions, as we have heard previously. More importantly, the ENSO phenomenon displays evidence of developing predictability over relatively long interseasonal time scales.

Recent studies have improved our knowledge of El Niño and its cool cousin La Nina, and have markedly improved prospects for understanding broad regional weather anomalies that will aid farmers and ranchers in their decisionmaking processes. Based on ENSO, scientists can predict with improving confidence whether the rainy season will be lengthened or shortened, or if precipitation differences will be significant.

The ENSO effect in the United States is generally weaker and interacts with significant natural variabilities. Crop yield relationships in El Niño or La Nina are more problematic, however. Seasonal rainfall is directly linked to ultimate crop yield, but there is also a critical issue of timeliness. Timeliness plus additional variables such as the threat of pests and frost, extreme temperatures, increased seasonal soil moisture reserves complicate the attempt to link yield to accurate seasonal or interannual forecasts.

Despite a weaker ENSO effect in problematic crop yield relationships, early studies show the potential for significant economic benefit to U.S. farmers if they have accurate ENSO forecasts. A soon to be published NOAA-sponsored study estimated the annual value of better ENSO predictions, as was cited earlier, range somewhere between $240 and $323 million per year, depending on the accuracy of the forecast. The higher value was estimated under the assumption that scientists could perfectly forecast ENSO events. The benefits of only the $323 million with a perfect forecast reflect the fact that even with this information farmers cannot avoid all losses.

Further research can identify a wider range of actions to mitigate weather-related damages associated with an El Niño year. For farmers and ranchers to realize the benefits of this research, it is critical that the information be made available in ways that are most useful to them.

USDA and its partner institutions understand the types of decisions farmers and ranchers must make and how they are affected by weather. More importantly, we have begun collaborating with NOAA about how to translate improved ENSO forecasts into information our customers can use to make better decisions.

Mr. Chairman, I would like to conclude my remarks by saying that USDA believes that an objective and timely analysis of the consequences of an El Niño phenomenon and the development of resource management strategies will help the agricultural industry maximize their benefits and minimize the cost. My testimony focuses on activities both underway and needed which are aimed at getting the most reliable weather and climate information collected, archived, and disseminated efficiently to the agricultural community for use in its decisionmaking processes.

When all is said and done, the farmer, rancher, forester, or commodity analyst must make the daily management decisions. Our goal is to provide the most accurate and timely information available. We appreciate the Subcommittee highlighting the importance of the El Niño phenomenon's impact on agriculture and your support for our efforts in this regard.

Thank you for the privilege of providing the testimony and I would be happy to answer any questions.

[The prepared statement of Mr. Gonzalez follows:]

STATEMENT OF DR. I. MILEY GONZALEZ
UNDER SECRETARY
RESEARCH, EDUCATION AND ECONOMICS
UNITED STATES DEPARTMENT OF AGRICULTURE
BEFORE THE
SUBCOMMITTEE ON ENERGY AND ENVIRONMENT
OF THE COMMITTEE ON SCIENCE
U.S. HOUSE OF REPRESENTATIVES
SEPTEMBER 11, 1997

Thank you Mr. Chairman for the opportunity to be here today to discuss El Nino and its affect on agriculture. I am also pleased to review with you how USDA and our partner institutions -- state cooperative extension services, land grant universities, and state agricultural experiment stations -- are focusing resources on this challenging phenomenon. With me today is Mr. Al Peterlin, the Chief Meteorologist for USDA, whom I may ask to help me respond to your questions.

Food and fiber production is a high-risk enterprise and the uncertainty of an El Nino phenomenon, which disrupts global rainfall and wind patterns, can make it even more so by playing havoc on agricultural production. Variation in weather is the major cause of variation in crop yield leading to variability in commodity prices and farm incomes. USDA and our partner institutions are also concerned about El Nino's affect on our natural resources. Research, economic analysis, and new knowledge dissemination about weather and climate variability, therefore, are critical to us in our day-to-day activities and how we serve our customers.

<u>CAUSE AND EFFECT: EL NINO AND AGRICULTURAL PRODUCTION</u>

In recent years, the U.S. National Oceanic and Atmospheric Administration's (NOAA) Office of

Global Programs has studied the multi-year cyclical interaction between the ocean and

atmosphere. Ocean waters in the tropical Pacific periodically increase in temperature to unusually

high levels and migrate eastward toward the South American coast. As the sea surface

temperature changes and interacts with the atmosphere above, atmospheric high and low pressure

centers are displaced, triggering unusually wet or dry regions over various land areas. This ocean

warming and eastward displacement is popularly called El Nino. Conversely, if the ocean

temperatures are colder than usual, a reverse phenomenon develops, called La Nina.

It is now fairly well recognized in scientific circles that the interaction of atmosphere and ocean,

primarily in the tropical Pacific, has a profound influence on global weather. This influence on

atmospheric circulation is manifest in many regions of the globe with the greatest and most

recognizable impacts found in Australia, South Africa, the west coast of South America, northern

Brazil, and the monsoonal areas of Asia, notably India and China. There is ample evidence in

published literature that El Nino/Southern Oscillation (ENSO) can influence significant shifts in

seasonal climate over some regions of the continental United States, especially along the west

coast and gulf coastal regions. More importantly, the ENSO phenomenon displays evidence of

developing predictability over relatively long (15 months) inter-seasonal time scales. [1]

---

[1]Rasmusson, E.M., and T. H. Carpenter. 1982: Variations in tropical sea surface
temperature and surface wind fields associated with the Southern Oscillation/El Nino.
Monthly Weather Review, 110, 354-383

Recent studies, many funded by NOAA, have improved our knowledge of El Nino and its cool cousin, La Nina, and have markedly improved prospects for understanding broad regional weather anomalies that will aid farmers in their decision processes. For Australia and parts of South America and Africa, where the ENSO signal is strong, there is evidence that forecasts of ENSO events led to significant improvement in agricultural performance. Knowing if the year would be wetter or drier, allowed farmers in these regions to plant crops suitable to the predicted weather and avoid the large swings in cereal production previously associated with ENSO events. In these regions of the world, seasons are divided into rainy and dry seasons. Based on ENSO, scientists can predict, with improving confidence, whether the rainy season will be lengthened or shortened, or if precipitation anomalies will be significant. The ENSO effect in the United States is generally weaker and interacts with significant natural variability. Thus, producing a reliable forecast of how weather will be affected in particular locations in the United States, given there is an ENSO event, is more difficult as is identifying specifically what farmers could do given the changed weather pattern.

Despite this weaker signal, early studies show the potential for significant economic benefits to U.S. farmers if they have accurate ENSO forecasts. A NOAA-sponsored study to be published in an upcoming issue of Climatic Change, a prominent journal dealing with climate issues, estimated that the annual value of better ENSO predictions ranged between $240 and $323 million per year depending on the accuracy of the forecast. The higher value was estimated under the assumption that scientists could perfectly forecast ENSO events. The benefits of only $323 million with a perfect forecast reflect the fact that even with this information farmers cannot avoid all losses.

USDA AND NOAA COLLABORATION

The NOAA-sponsored study on the potential value of better forecasts of ENSO events has important implications for agricultural research, education, and extension. The study assumed that farmers understood the accuracy of the forecasts and responded ideally to the forecast information. To begin to realize these benefits it is critical that the information be made available to farmers in ways that are most useful to them. USDA agricultural research, education, and extension agencies understand the types of decisions farmers and ranchers must make and how they are affected by weather. More importantly, we have begun collaborating with NOAA about how to translate improved ENSO forecasts into information our customers can use to make better decisions. Specific partnerships - to name a few - are:

- The Office of the Chief Economist's World Agricultural Outlook Board (WAOB) monitors world weather and climate through its Joint Agricultural Weather Facility to assess the impact of all weather and climate events on global crop production. The WAOB prepares and releases its World Agricultural Supply and Demand Estimates reports and keeps the Secretary and other USDA officials informed of national and international weather conditions which may impact production, prices, and trade.

- The U.S. Forest Service, in collaboration with NOAA's National Weather Service (NWS) and the U.S. Department of Interior, is intimately involved with weather and climate monitoring in the National Forests through its Remote Automated Weather Stations network and daily operational forecasting in support of controlled burns and fighting wildfires;

- USDA and State Agriculture Experiment Stations (SAES) serve NWS in their

Cooperative Weather Observer Network. Many Experiment Stations have over 100 years of observations and represent a high proportion of the approximately 20 Climatological Reference Stations that provide a wider array of long-term agricultural weather observations to NWS (including soil temperature, evaporation, solar radiation);

- The Agricultural Research Service (ARS) and NOAA are working together to make weather radar data more useful to the agricultural community. Research is needed to translate radar information into accurate spatial and temporal rainfall distributions. ARS has unique ground based experimental facilities for developing and validating new and improved algorithms for enhancing the performance and reliability of rainfall estimates based on the radar data;

- The Economic Research Service is collaborating with NOAA on the impacts of weather variability on agriculture including developing a better understanding of the ability of farmers to detect and respond to changing climate given the significant normal variability in weather.

RESEARCH: PLANNING AND MITIGATION

Further USDA research can identify a wider range of actions to mitigate weather-related damages that could be taken with better forecasts, thus avoiding a larger share of the damages associated with an El Nino year. The NOAA study considered only a limited set of options for mitigating weather-related losses under ENSO events--principally changing the crop mix. Some of the more promising adaptations may involve other changes in farming practices. For example: farmers anticipating a strong likelihood of less than normal rainfall might use caution in applying high rates of fertilizer which may not be fully used by crops because of drought stresses. Given sufficient lead time, farmers in crop transitional areas might decide to switch to crops better able to withstand "expected" anomalous weather conditions. Farmers anticipating a dry season could switch to a drought resistant crop such as sorghum vs. planting a more moisture sensitive corn crop or vice versa. Ranch and dairy operators, anticipating either a dryer or more moist grazing season, could plan to purchase more or less hay or supplemental feed, plan for increased or decreased time on pasture, or cull or increase herd numbers. The Forest Service could perhaps better respond to the threat of wildfires if an El Nino or La Nina based forecast called for a higher risk during the fire season in the west, or in the southeast.

Crop yield relationships and El Nino or La Nina are more problematic. Seasonal rainfall is directly linked to ultimate crop yield, but there is also a crucial issue of timeliness. Timeliness, and a whole host of additional variables including pest threat, local geographic anomalies in extreme temperatures, frost threat, pre-seasonal soil moisture reserves and others, complicate the attempt to link yield, even to accurate seasonal or interannual forecasts. Examples of uses of seasonal and interannual forecasts are:

- Prediction of harvest dates, especially in crops where harvesters must schedule equipment and processors must optimize input into production facilities; and

- Regional and global pest outbreaks, such as rust on wheat or locusts on cereals, can be predicted better with knowledge of seasonal forecasts.

CONCLUSION

Mr. Chairman, I would like to conclude my testimony by identifying 2 new weather-related efforts that we believe will help improve the performance of U.S. agriculture:

1. Responding to the devastating drought in the Southwest in 1996 that was related to the La Nina event, the Western Governors' Association (WGA) issued a Drought Report in November 1996. The report includes a number of recommendations to improve federal and state responses to drought, and emphasized the need for incorporating mitigation and preparedness measures in governmental drought programs. One key recommendation called for the creation of a regional drought policy and coordination council. As a result, a Memorandum of Understanding (MOU) was signed in early 1997 by WGA, USDA, Department of the Interior, Department of Commerce, Federal Emergency Management Agency, Small Business Administration, Fort Belknap Community Council, Chickasaw Nation, and the National Association of Counties. The MOU established the Western Drought Coordination Council (WDCC) to work toward improving drought response in the west and recognized USDA as the lead federal agency for drought.

New Mexico Governor Gary Johnson and Deputy Secretary of Agriculture Richard Rominger co-chair the WDCC. The Council's objectives are to: (1) assist state, local, and tribal governments develop drought preparedness and mitigation programs; (2) improve coordination and information exchange at all levels of government; (3) improve drought response of federal, state, and local programs; and (4) heighten awareness of drought management issues and promote efficient use of water in the west.

2. The agricultural research and extension bill reported out of the Senate Agriculture Committee on July 30, 1997 included a provision establishing the National Agricultural Weather Information

System (NAWIS). With respect to El Nino, this new authority will allow USDA and partner institutions to conduct research at the regional level related to agricultural impacts from El Nino and other weather anomalies, and improve interaction between NWS, USDA, and state climatologists who will best be able to interpret El Nino impacts in local areas. NAWIS can also serve as the focal point for day to day management of the drought activities being accomplished by the WDCC.

USDA believes that an objective and timely analysis of the consequences of the El Nino phenomenon and the development of resource management strategies will help the agricultural industry maximize their benefits and minimize the costs.   My comments today have focused on the activities, both underway and needed, which are aimed at getting the most reliable weather and climate information collected, archived and most importantly, disseminated efficiently to the agricultural community for use in their decision making processes. When all is said and done, the farmer, rancher, forester, or commodity analyst must make the day-to-day management decisions. Our goal is to provide the most accurate and timely information available. We appreciate this subcommittee highlighting the importance of the El Nino phenomenon's impact on agriculture and your support of our efforts in this regard.

Thank you again for the privilege of submitting this testimony. I will be pleased to answer any questions you may have at this time.

I. MILEY GONZALEZ, Ph.D.
Biographical Sketch

Dr. I. Miley Gonzalez was born in Ysleta, Texas, on July 30, 1946. He attended the University of Arizona where he received both a Bachelor's and Master's Degree in Agricultural Education. He completed his doctoral work in Agricultural and Extension Education at Pennsylvania State University in 1982 while serving there as a state 4-H specialist for two years. He subsequently was appointed assistant professor in Agricultural and Extension Education at Iowa State University and went on to serve as assistant director for International Agriculture Programs from 1988-1991.

He first joined New Mexico State University in 1991 as professor and head of the Department of Agricultural and Extension Education. In 1994, he was appointed assistant dean and deputy director of the Cooperative Extension Service. For the past year, he has served as associate dean and director of academic programs in the College of Agriculture and Home Economics.

Dr. Gonzalez has conducted over 100 seminars and workshops for more than 800 extension workers and vocational educators in Arizona, Iowa, Pennsylvania, and several Latin American countries. His teaching has focused on problem-solving techniques, program planning and evaluation, curriculum development, and beginning teacher and extension agent training.

From 1987 to 1989, he conducted workshops for over 100 agriculture and home economics teachers and 40 university faculty members from Central American institutions. He has served on numerous evaluation and task force teams at the local, national, and international levels. His research and scholarly activities have focused on the areas of international development, extension education, adult and non-formal education, technology transfer, and leadership development and training.

Dr. Gonzalez has received numerous awards and honors from various professional groups, teachers, institutions and many other organizations such as 4-H and the FFA. He earned the American Farmer Degree in FFA and has received the Distinguished Service Award from the American Association of Agricultural Education for both domestic and international work in the western region.

Dr. Gonzalez has a background in agricultural production, marketing, and processing. He was reared on a small farming and ranching operation in southeastern Arizona and managed farming and marketing operations in northern Mexico. He and his wife, Julie, have five children and one granddaughter.

Chairman CALVERT. Thank you, Doctor.
Mr. Wheeler?

TESTIMONY OF DOUGLAS WHEELER, SECRETARY, CALIFORNIA STATE RESOURCES AGENCY, SACRAMENTO, CA

Mr. WHEELER. Good morning, Mr. Chairman.

I am Doug Wheeler, Governor Wilson's Secretary for Resources, who has responsibility for California's natural resources programs. I am accompanied this morning by Scott Cameron of the Governor's Washington office. And I thank you for this opportunity to participate and to think through with the Committee the need to take the predictive information—about which we have heard a lot this morning—and apply it to the problems we confront as resource managers and as officials at the state and local level, and to better disseminate the information that might be helpful to the citizens of California as they prepare for this extraordinary event.

It is significant to note that one of our departments, the Department of Boating and Waterways, in conjunction with Scripps, conducted the first public meeting on this subject, in August. As you heard from Dr. Barnett this morning, it was very well attended. There is a high degree of interest in these issues in California and a high degree of preparedness.

The numbers you have heard discussed are variable. The impacts regionally within our State are expected to vary substantially over that 1,100 mile stretch of California coastline. So it is difficult to predict with absolute certainty, as you heard this morning, just what those impacts will be and where they will occur. But I do have for you a very brief summary of our best estimate at present of the effects and their regional impacts in California ranging from north to south.

In the northern part of the State, along the north coast, we expect a dry to normal rainy season, which would also include high tides and waves and river levels somewhat lower than usual. In the central valley of California, we expect a normal to wet season, where the status of levies, of course, and the need for accurate river forecasts, is increasingly important.

In the San Francisco Bay and the delta of the San Joaquin and Sacramento Rivers, we expect also normal to wet conditions, that is, somewhat higher than usual rainfall. We are concerned there about high tides and waves, particularly in the San Francisco Bay region, and about the security of our levies in the delta.

In the southern Sierra, particularly in the Tulare Lake Basin, we expect normal to wet conditions, wetter than usual. And we could anticipate from that flooding of the Tulare Lake Basin and impacts upon that agricultural community.

In southern California, we expect the most severe impacts, as you have already heard this morning. And along the south coast, Mr. Chairman, extraordinarily wet conditions accompanied by high tides and waves. This is what did occur during the 1982 and 1983 season. And as you have already heard this morning, that is the closest analogy to our experience in 1997.

I think to paraphrase Dr. Barnett, we are preparing for the worst in California and hoping for the best. This situation, although not exactly similar in character since 1982–1983, has its parallel in a phenomenal series of natural occurrences which have beset California in the last 8 or so years, but most recently with the floods of 1997. I think, although it is regrettable to have to say this, California is becoming accustomed to dealing with extraordinary natural phenomena. And the people of California, as you know, Mr. Chairman, have developed remarkable resilience in the face of those extraordinary events. So, too, we are preparing for El Niño, expecting that it could be quite bad but hoping that it won't be.

The 1997 floods were the most extensive in our history. You know that there were lives lost. You know that there was nearly $2 billion in property damage. I served as Chairman of the Governor's Flood Emergency Action Team during that period and we have made, coming out of that experience, some 50 different recommendations now being implemented by local government, by the State of California, and by our federal partners to guard against the flooding consequences of El Niño.

But I want to emphasize that the flooding, which is likely to occur, occurs in a part of the State where, in the case of the Central Valley, we are quite well prepared as a result of our response to the 1997 episode, but in southern California not so well prepared because we don't have the same systemic approach. Therefore, I think it is important to focus our efforts at southern California, where we can expect the most severe impact.

The State government's response to El Niño in anticipation of these effects, particularly in southern California, takes several forms. I will just summarize those for you because my testimony includes a more complete account. Again, we are moving from the predictive phase to the dissemination phase and the application phase. We think it is very important that the public in California be well informed about this phenomenon and about its likely impact.

For that reason, we have established a web site on California's natural resources database. It is called—and I invite your attention to—the California Environmental Resources Evaluation System, CERES.CA.GOV/El Niño, where the best available information—local, state, and federal—is already available to everyone on the Internet through that mechanism. At the same time, we are going to be working with the Governor's Office of Emergency Services to provide that information to local governments and to private citizens who could make use of it, in cooperation with our colleagues at FEMA.

In addition to the need for information dissemination and public awareness, we are very much concerned with emergency preparedness. Again, we have had opportunity to hone our skills in this regard as the result of fires and floods and earthquakes over the last few years. That system (the Standard Emergency Management System) will be in place and will be strengthened in anticipation of El Niño. That is a partnership focused on the incident command approach in which resources of all levels of government are pooled to meet the needs of a specific region in the event of an emergency.

We are concerned with the protection of public infrastructure, for which it is government's responsibility to protect, to manage. That is particularly the case with respect to our energy grid, both the generation of electricity and its distribution in California. We want to maintain the integrity of our state and federal highway system, which is a vital link for California's economy and public safety. And we are of course concerned about coastal installations which are owned or managed by the state government as part of our ongoing responsibility.

Finally, we are going to emphasize technical assistance to local governments. We learned during the floods in January of 1997 that the state and the Federal Government had resources to offer which were not as often being delivered to the place where they are needed most by reason of a failure to coordinate more actively and to focus those efforts. We are now anticipating that need and preparing for it by the deployment of resources early rather than late in our relationship with local government.

Let me say that—as I mentioned a couple of times already—this is a partnership effort, at government level, and that nothing we can do as a State can be done without the assistance of local government, and particularly without the assistance of our colleagues in the federal establishment. But we worked very well together during the flood response to curtail the impacts of those most extensive floods ever in California's history in January, as you know. And we are calling upon our federal colleagues to increase that assistance and that interaction now as we confront the prospects of El Niño.

While although Mr. Rohrabacher may welcome El Niño for the positive effects, I have to say that as someone concerned with responding to it this winter, I am not looking forward to it, exactly. But we are confident that the people of California, given adequate preparation, given the resources the Governor will bring to bear, will once again demonstrate the resilience which has enabled us to survive these events. The Governor has said on many occasions— unfortunately had to say on many occasions—that the worst in weather brings out the best in Californians. We think that will be true again this year.

[The prepared statement of Mr. Wheeler follows:]

Testimony of

Douglas P. Wheeler
Secretary for Resources
State of California

Before the Subcommittee on Energy and Environment
of the U.S. House of Representatives
Committee on Science

California Preparations for El Niño

September 11, 1997

Mr. Chairman and Members of the Subcommittee, I am Douglas P. Wheeler, Governor Wilson's Secretary for Resources.

Thank you for the opportunity to speak today on the activities California is undertaking in preparation for El Niño. I will provide a summary of ongoing and proposed actions as well as provide a background of California's experience in dealing with large-scale natural disasters.

Background

A large El Niño event is underway and forecasted to continue into the winter. Though we cannot be sure, we are told to expect substantially more rainfall than normal, especially in the central and south coastal areas of California. Fall rains are likely to come earlier, and if the early rains are sufficient to soak the ground, we could have an early and long flood season. Based on past history, the flood risk would be higher in the coastal and lower elevation terrain and not as high on major Central Valley rivers arising in the Sierra.

Higher ocean levels are expected, one-third to one-half foot above normal. Given that peak tides are highest in the winter, and since the highest tides of winter always occur in the morning, storm preparations will have to be carried out at night. Coastal damage and erosion are most severe when storm waves coincide with peak high tides. Potential coastal impacts include: structural damage from wave overtopping, onshore flooding and wave impact, and land loss due to cliff failure and beach erosion.

Public lands and infrastructure as well as private property are at risk. State Beaches are some of the most vulnerable, with bulkheads, seawalls, parking lots and rest rooms in danger. Boating facilities and boats are also at risk. During the winter of 1982-83, a strong El Niño event, over $100 million of damage occurred to public and private coastal infrastructure. One third of the damage (about $35 million) occurred to public recreational facilities along the coastline in Central and Southern California. Farther north, while the Central Valley flood control system performed well considering the record rainfall and runoff, ten Delta islands were flooded due to levee failures.

Impacts to inland watersheds could be highly variable. The Department of Fish and Game notes that the impact to California's inland rivers and streams will depend upon the condition of the watersheds. For watersheds that are stable, the high volume of rainfall and runoff that frequently accompanies an El Niño may wash sediment out of the gravel, thereby improving spawning conditions for fish. For unstable watersheds, El Niño may pose a more severe threat, as it can cause hillsides to collapse into the river or stream.

State Preparedness

Given the great uncertainty of what will actually occur, California will be prepared to meet any foreseeable threat. Regrettably, California has an enormous amount of experience with natural disasters, most recently the floods of 1997, and has learned to improve its response in the wake of each such catastrophe.

In January 1997, California was struck with the most extensive flooding in the State's history, affecting thousands of residents and causing nearly $2 billion in damage. At the height of our winter storms, on January 10, Governor Wilson established the Flood Emergency Action Team (FEAT) to coordinate the State's emergency response and recommend long-term actions to protect Californians from future flood disasters. I served as Chair of the FEAT. In its final report, the FEAT made more than fifty recommendations that now are the basis of the State's flood policy and serve as a template for all of the State's emergency response systems.

While the State regularly undertakes emergency preparedness exercises and activities, the following additional recommendations by the FEAT, which are currently being implemented by various State agencies, are particularly relevant to the State's preparation for El Niño.

<u>Improving emergency response coordination and operation</u>: In conjunction with federal and local agencies, the Governor's Office of Emergency Services is developing and testing guidelines for a Standardized Emergency Management System that will clarify and coordinate all joint emergency field operations. The Office of Emergency Services and the Department of Water Resources are conducting flood emergency workshops and have initiated intensive training for local agencies aimed at improving communication activities during disasters.

In addition, the Office of Emergency Services is currently developing, in conjunction with other State agencies and the U.S. Coast Guard, procedures for the emergency closure of Sacramento-San Joaquin Delta waterways in flood conditions.

<u>Improving and expanding existing flood data</u>: The California Department of Water Resources has installed nearly 50 stream gauging sites in areas that have high flood probability to expand and enhance the State's existing flood data. Earlier this year, the State also called on the U.S. Geological Survey to expand its surface water data collection program so that we might obtain better, early data about flood threats. Unfortunately, the federal government indicated that such expansion was unlikely due to budget limitations.

In addition, all State agencies have been reviewing on-going activities in anticipation of El Niño:

- California was one of first entities to provide comprehensive information on El Niño on the Internet. CERES, the California Environmental Resources Evaluation System (http://www.ceres.ca.gov), one of the world's largest environmental web sites, was the first to distribute information to the public on a major disaster during the floods of 1995. CERES has continually provided up to date information on El Niño, preparedness, and other related topics from government, media, nonprofit, and private organizations. CERES, which is a ready and viable source for the dissemination of information, could be a critical communication tool in disaster situations;

- To ensure that there is thorough and effective coordination and communication among the relevant State Agencies, and with their local counterparts, El Niño preparation is now a Cabinet-level priority;

- Affected State agencies are preparing personnel to staff coordinated emergency operations centers, and provide emergency response in the event of an El Niño emergency. State Agencies are also aggressively undertaking preventive maintenance of facilities that will minimize the effects of natural disasters such as flooding; and

- The California Energy Commission is taking special precautions to ensure that the State's electrical generation and distribution systems are "disaster-proof".

Preparation for El Niño, and the potential effects it might unleash, are one of the highest current priorities of the Wilson Administration.

California will continue to provide the public with the best available information through the internet, media briefings, workshops, and other avenues of communication. California will also continue its efforts to ensure that its Office of Emergency Services and other State agencies are prepared for El Niño and its potential effects. Finally, the State will continue its outreach and training for local governments and organizations to ensure that there is sufficient coordination and communication between the State and federal governments and local communities.

However, there remains an urgent need for better coordination within the federal agencies, and between them and California, as well as the commitment of federal funding for critical activities and initiatives related to El Niño, if California, and the Pacific Coast region are to be as well prepared as possible. I call upon the subcommittee to highlight the Federal role in preparing for El Niño, without which continuing partnership we will not be adequately equipped to deal with its effects. Specifically:

- We request that the Army Corps of Engineers use its "PL 84-99 Advance Measures" authority for those portions of the Sacramento/San Joaquin Flood Control System damaged in the January 1997 floods and not yet repaired;

- We request that the U.S. Geological Survey expand its surface water data collection program so that we might obtain better, early data about flood threats;

- We request continued cooperation from the Federal Emergency Management Agency and the Department of Defense in collecting extremely accurate digital elevation map products of California's Central Valley, so we can better plan for and react to any flooding that occurs this winter in that region.

- We urge continued coordination from NOAA in making the most accurate and recent information available for integration into the CERES emergency information system. We suggest developing a method of email communication with CERES and others that would describe NOAA updates, new links, and future offerings to help us highlight their information on the CERES El Niño page (http://ceres.ca.gov/elnino);

- We urge support of making NASA data available in a usable format via the internet for disaster preparedness, disaster response, and disaster recovery. It would then be integrated into the CERES emergency information system;

- We request that emergency situations be given a high priority for military or NASA missions to gather digital high altitude photographs to assist the State of California in assessing natural disasters;

- We also request support from the Federal Communications Commission to interface the Emergency Alert System with information the State of California provides to the public via the internet.

Thank you for the opportunity to speak today. I will now answer any questions you might have.

The Resources Agency

Pete Wilson
Governor

Douglas P. Wheeler
Secretary

of California

California Conservation Corps • Department of Boating & Waterways • Department of Conservation
Department of Fish & Game • Department of Forestry & Fire Protection • Department of Parks & Recreation • Department of Water Resources

Doug Wheeler is California's seventh Secretary for Resources, having been appointed to this cabinet position by Governor Pete Wilson shortly after his election in November 1990. Wheeler was sworn into office on January 7, 1991.

As Secretary for Resources, Wheeler is responsible for the resource management programs of 18 departments, conservancies, boards and commissions with combined budgets of nearly $2 billion and total staff of approximately 13,000. These entities include the Departments of Water Resources, Fish & Game, Forestry & Fire Protection, Parks & Recreation, and Conservation, and the California Coastal Commission.

Since the adoption by Governor Wilson of his "Resourceful California" agenda in early 1991, Wheeler has been instrumental in the design and implementation of nationally-recognized policies and programs that seek to integrate economic and environmental objectives and to advance "ecosystem management." Among the best known of these are the Natural Community Conservation Planning (NCCP) program in Southern California, the State-Federal CALFED Bay-Delta Program for the San Francisco Bay/Sacramento-San Joaquin Delta estuary, and the on-line California Environmental Resources Evaluation System (CERES).

In addition to his administrative responsibilities, Wheeler serves as chair of the Governor's inter-agency Water Policy Council and the Flood Plain Management Task Force, and co-chair of the State-Federal CALFED Bay-Delta Program and the Governor's Wetlands Task Force. He is also chair of the California Biodiversity Council, which has 36 member agencies from local, State, and Federal government.

In 1996, Wheeler received the President's Award for Sustainable Development.

The Resources Building   Sacramento, CA 95814   (916) 653-5656   FAX (916) 653-8102

California Coastal Commission • California Tahoe Conservancy • Colorado River Board of California
Energy Resources, Conservation & Development Commission • San Francisco Bay Conservation & Development Commission
State Coastal Conservancy • State Lands Commission • State Reclamation Board

A long-time leader in the field of conservation, Wheeler served as Vice President of the World Wildlife Fund and the Conservation Foundation in 1990, having served as Executive Vice President and Vice President of the Conservation Foundation from 1987-1990. He was Executive Director of the Sierra Club (1985-87), Founder and President of the American Farmland Trust (1980-85), Executive Vice President of the National Trust for Historic Preservation (1977-80), Deputy Assistant Secretary of the Interior (1972-77), and Legislative Counsel and Legislative Attorney for the Department of the Interior (1969-71).

Wheeler is a member of the Board of the California Nature Conservancy and a member of the Board of Visitors of the Duke University School of Law.

Doug and Heather Wheeler are the parents of two sons and reside in Yolo County. He is a graduate of Hamilton College (1963) and the Duke University School of Law (1966), and a member of the North Carolina Bar Association.

MAINTENANCE OF FLOOD CHANNELS IN CALIFORNIA

Chairman CALVERT. I thank you for your testimony, Mr. Wheeler. And I am grateful that you are involved in the planning on the worst and hoping for the best. California certainly has had its unfortunate days in the past.

One of the issues that concerns me—and I know Mr. Brown and I share the same area in southern California—you mentioned that southern California, based upon these predictions of El Niño, is the area that could be the most affected as the area around the Santa Ana River. Though we are working toward the All River Plan and improving the Pravo Dam Basin area and the Seven Oaks Dam is under construction, as you know, those projects are not yet complete. In the meantime, urbanization has continued to take place down river of the Santa Ana River since 1982–1983, which was a large flood event, as you know. And since then, a lot—it seems to me—maintenance and flood channels and ditches, and certainly in the Santa Ana River, could potentially cause some problems.

Are you looking into that or working with the Corps of Engineers to see if there is some maintenance that could be done preemptively to mitigate for that?

Mr. WHEELER. You make a good point. There was a lot of discussion about this during the floods of the winter. But those floods, has you have already indicated, had their effect in the Sacramento and San Joaquin Basins more than in southern California. We now anticipate a comparable effect in southern California. And this issue of the maintenance of our floodways and stream channels is one to which both the state and Federal Government have devoted a lot of attention.

I can assure you, Mr. Chairman, that to the extent permitting is required and the processing of permits to assure clearance—those needs will be anticipated and expedited. That was true in northern California in the winter. It is now understood by the state government and the Federal Government, I am sure, to be a high priority. We will turn our efforts to making sure that the system in place is effectively operated as possible.

Chairman CALVERT. I am very hopeful that will occur. As you know, Marieta Creek in my district I share with Ron Packard certainly had severe flood damage just a few years ago. We are hoping that the maintenance, working with the State and the Corps of Engineers and other agencies, to make sure that that maintenance and improvements take place now.

Mr. WHEELER. Mr. Chairman, not to appear ungratefully or overly ambitious, I have included in my testimony a list of specific items that we had hoped the Federal Government would provide in helping us to provide for El Niño. One of those is to ask the Corps of Engineers for advanced deployment of its resources so that we are anticipating the problems that you describe and not responding to them after the fact when it is obviously much more difficult to mount an effective response.

Chairman CALVERT. Very good.

Mr. Roemer?

DISASTER RELIEF SUPPLEMENTAL APPROPRIATION

Mr. ROEMER. Mr. Armstrong, with relation to FEMA, one of the things I said to the last panel was that we have certainly sent a lot of money out from Washington, DC. to different regions of the United States for floods, hurricanes, tornadoes, fires, El Niño, and so forth and so on. Should we be anticipating another federal disaster supplemental with this kind of prognosis and these kinds of things we are thinking about today?

Mr. ARMSTRONG. Unfortunately, we are in a climate where I guess you always have to anticipate another disaster supplemental. If things happen the way they have happened in the past, we would anticipate that there could be widespread flooding in California, amongst other jurisdictions. The good news is that we are moving along at a fairly good clip at expediting mitigation efforts to try to position structures so that they can be better able to withstand flooding hazards, in particular. Just with the recent flooding—the 1997 floods alone—91 percent of projects proposed for elevation and 100 percent of the projects proposed for acquisition have already gone through the process just after the flood of September.

So we are trying as much as possible to promote mitigation. We are trying to promote public education. We are trying to get people to buy flood insurance. All of these things, if they are implemented, will reduce the impact on the Federal Treasury.

SAVINGS FROM EL NIÑO PREDICTIONS

Mr. ROEMER. Do you have any estimate as to how much that saves when we do the education and the mitigation and the proactive activities that Mr. Wheeler is talking about?

Mr. ARMSTRONG. All we can do is point to previous incidents of flooding in other parts of the country as anecdotal information. For example, in the aftermath of the Mississippi floods, the floods of 1993, we have determined that for every $1 spent we are saving $2 when we involve ourselves in mitigation projects.

DISASTER RELIEF SUPPLEMENTAL APPROPRIATIONS

Mr. ROEMER. Do you have an estimate over the last decade of the amount of money that we have spent out of the Federal Treasury for disaster assistance supplementals?

Mr. ARMSTRONG. Mr. Adamchek is with me from our Response and Recovery Directorate. We will provide it for the record.

Mr. ROEMER. Do you have any kind of a rough guess per year, going back for the last decade, as to what the variety has been from year to year, from $1 billion to several billion in disaster relief supplemental bills we have concerned in Congress?

Mr. ARMSTRONG. $1.2 billion.

Mr. ROEMER. A year?

Mr. ARMSTRONG. Yes.

Mr. ROEMER. About roughly $1.2 billion a year that we have appropriated through this Committee out of federal treasuries to go back to different regional flooding, fire, El Niño areas?

Mr. ARMSTRONG. That is absent the Northridge earthquake, but that is a per year average.

MITIGATION IN CALIFORNIA

Mr. ROEMER. Do you have an estimate for California?

Mr. ARMSTRONG. A lot of it.

[Laughter.]

Mr. WHEELER. Not enough.

Mr. ROEMER. Not enough and a lot. My taxpayers in Indiana would probably say a lot, and your taxpayers in California would probably say not enough.

Mr. ARMSTRONG. I would add, though, if I could, that just in our interaction with the State of California in the past several years, we have seen a huge movement forward with our relationship in terms of expediting mitigation projects and coming up with some new strategies to promote mitigation in the State. I think we are much better positioned today than we have been in the past.

COSTS OF MITIGATION

Mr. ROEMER. So we are not only spending more now on California for the disaster relief after it has happened, we are spending more out there for mitigation as well?

Mr. ARMSTRONG. Well, I hope we are spending more on mitigation and less on response and recovery. In the long term, if we spend more on mitigation we will reduce the overall expenditures of the Federal Government.

DISASTER RELIEF SUPPLEMENTAL APPROPRIATIONS

Mr. ROEMER. Do you have an overall suggestion, Mr. Armstrong—and I am not trying to put you on the spot at all, your answers have been very helpful to me—do you have an overall perspective as to what we should do in Congress, rather than address these emergency disasters year by year as they come at us in different regions of the country? I know there have been different proposals before Congress as to what we should do differently rather than across the board cuts, rather than going into debt. Do you have any suggestions?

MITIGATION

Mr. ARMSTRONG. Well, not to sound like a broken record, but both the Director and the Administration have proposed to the Congress that we do look more at pre-disaster mitigation. We really believe that pre-disaster mitigation is the old adage of an ounce of prevention is worth a pound of cure. The problem is that these involve local land use decisions and it involves the active participation of municipal and county governments as well as state legislatures. There are competing issues when you get down to the local level in terms of land use.

But as has been said by many of the other witnesses today, we can't control the climate. But we can control and respond to how we choose to live with that climate. The kind of land use decisions that are made in terms of putting at-risk structures, such as mobile home parks, in flood plains, not insuring public buildings and then having to come in as a Federal Government and pay for the rehabilitation or restoration of those buildings when those buildings should have been insured, encouraging States to have their own mitigation funds and their own contingency funds, promoting a better sense of education within the construction and building industries in terms of the types of materials that could and should be employed.

We are working with a variety of organizations to promote a multi-hazard approach to building codes so that when someone, for example, constructs a building in the gulf coast they are not just looking at hurricanes but they are looking at floods in how they build that building. Or in California they are looking not only at the seismic threat but the fire threat as well. I think that everybody up and down the chain—and not only in the Federal Government, but in the non-profit and private sectors—needs to look at opportunities for mitigation.

I think that is beginning to take hold with folks. The American Red Cross is now promoting mitigation. We are working closely with the Institute of Business and Home Safety, which is a consortium of insurance companies across the country to promote mitigation. We have a variety of earthquake consortia across the country that are promoting mitigation. Our new disaster-resistant community initiative is designed to showcase communities where there are already public-private partnerships occurring and a local climate of political support to encourage mitigation activities. So we can spotlight these communities and show other cities and other counties that mitigation does work.

Mr. ROEMER. You were certainly prepared for that one. You had a good answer and I appreciate your answer.

[Laughter.]

Chairman CALVERT. Thank you, Mr. Roemer.

Mr. Ehlers?

SELF-INSURANCE FUNDS

Mr. EHLERS. Thank you, Mr. Chairman.

First of all, let me follow the same line of questioning because this is one of the concerns I have. And I appreciate the answer you just gave because I think that is absolutely crucial that the Federal Government take that path and persuade the States and local governments and the private sector to follow that path as well. I think it is absurd with the present system where we just let it happen and then say, "Oh my, you need some money," and then we send some money.

A corollary to that, something I believe is very important in the Federal Government—and I would appreciate your response to this as to whether you think it is workable and whether it should be done—it bothers me that we handle this through supplemental appropriations, which takes some time, it is hard to gauge precise needs, sometimes we may appropriate too much and the money is perhaps not always wisely used because of the urgency.

It seems to me that it would be a much better approach to do what we used to do in local government, since we didn't have any resources for getting additional money. We not only had contingency funds. We went beyond that and basically set up self-insurance funds using standard insurance practices. I believe it would be appropriate for FEMA to do that with the requisite appropriations from the Federal Government and to establish a self-insurance fund with all the standard practices of setting up appropriate reserves, investing appropriately, and insuring that the money is there when the need arises, but also having very firm guidelines as to how it is to be expended, and for what purposes, and which people qualify. If they have followed proper practices, they qualify. If they haven't, they don't.

What is your opinion of the possibility of setting something like that up?

Mr. ARMSTRONG. Well, I won't even pretend to be an expert on insurance. It is our directive—the mitigation directive works on compliance with flood plain mapping that has a direct relationship or corollary to insurance. The Federal Insurance Administration is in a different part of FEMA. But I will say, as I said earlier, that a greater use of insurance will reduce the cost to the Federal Treasury.

In addition, I believe that the cost of insurance will also act as an incentive for individuals, property owners, business persons to make smarter choices on where and how they locate. Those smarter choices will also reduce the impact on the Treasury.

Mr. EHLERS. I am basically talking about a double-barrel approach. First of all, that the Federal Government's—all the emergency management funds be set up in a self-insurance fund, and secondly that you work with the private sector to persuade individuals to purchase insurance, which may have to be to some extent subsidized by you.

But I think that is a useful avenue to pursue, and I would appreciate it if you would pass that on to your colleagues.

Mr. ARMSTRONG. Certainly. I think the folks at the Federal Insurance Administration (FIA) would be glad to discuss it with you.

Mr. EHLERS. Thank you very much, and I yield back.

Chairman CALVERT. Thank you, Mr. Ehlers.

Ms. Hooley?

Ms. HOOLEY. Thank you, Mr. Chairman.

I regret that I wasn't here for the earlier panel, but a lot of times you have conflicts, so I appreciate this opportunity to ask a question and I would also like to submit a statement for the record, if I may.

Chairman CALVERT. Without objection.

[The prepared statement of Ms. Hooley follows:]

Opening Remarks
Honorable Darlene Hooley
Hearing on Preparing for El Niño
Committee on Science
September 10, 1997

I would like to thank the Chairman for holding this hearing on an issue that we've all heard a great deal about. To a great extent, the reports have been quite extreme, and a number of people have approached me with some dire predictions. So I see this as a wonderful opportunity to learn more about the science behind the climate forecasting and the impact estimates that have been attributed to the so-called "El Niño."

From what I've heard in the press, it appears that we're in for a year of relatively bizarre weather: flooding in some parts of the country, droughts in other parts, and perhaps a more active hurricane season. In speaking with climatologists in my state, it appears that there may be some misconceptions about what the phenomenon is and what its specific effects will be, and they are quick to add that we still have a great deal to learn.

Many scientists have noted the periodic occurrences of these events and suggested that abnormally large occurrences may take place every ten to fifteen years. The 1982-1983 event was the strongest on record, leading to thousands of deaths and devastating property damage. However, that event is reportedly somewhat different than the current El Niño. The 82-83 event peaked during the late winter, while this one seems to have peaked during the summer, and other indicators are slightly different as well. In the Pacific Northwest, this El Niño may produce powerful storms, but not the extreme flooding that we have experienced in recent winters.

However, it appears that these climate changes may have a significant impact on the Pacific Northwest salmon population. There is increasing evidence that salmon populations in the northeast Pacific are significantly influenced by long-term climate changes. It appears that the salmon show behavior that very closely follows the climate cycles. For example, spring chinook tend to decline during the warm, dry periods associated with an El Niño event and return during the cool, wet periods. Obviously, these stock patterns have a significant impact on the fishing industry and the economy of the Northwest.

The ability to accurately predict such weather events is critical to key decision-making processes in this industry, as well as other agricultural and even energy related industries. But to make such predictions, we must have well-established cause and effect relationships and have developed strong indices using ocean temperatures, air temperatures and oscillation data. I look forward to hearing more today about our current ability to quantify certain effects, and make high-probability predictions.

Again, I thank the Chairman for holding this hearing and will not further delay the anticipated testimony of our witnesses.

EL NIÑO AND SALMON IN THE NORTHWEST

Ms. HOOLEY. Thank you.

I have a question for Mr. Gonzalez.

It appears that these climate changes will have a significant impact on the Pacific Northwest salmon population, and there has been a lot of talk about salmon in our State. And there is increasing evidence that salmon populations in the north Pacific are significantly influence by long-term climate changes. They show patterns that very closely relate to the climate cycle.

For example, I know the Spring Chinook tend to decline during the warm, dry periods associated with an El Niño, and then return during the cool, wet periods. Obviously these stock patterns have a significant impact on the fishing industries and the economy of the northwest.

So my question is, What data is needed to provide the reliable weather and climate information so that the northwest fishing industries and other agricultural industries can make a proactive decision and try to avoid these severe economic impacts?

Mr. GONZALEZ. Thank you for the question.

As I indicated earlier—and I am not sure if you were here at that moment—not being a scientist in this area——

Ms. HOOLEY. I heard that part.

Mr. GONZALEZ. In dealing with some of these things, I do know that we have I think as part of our ARS research capability and in our partnering with land grant universities and certainly with the other federal agencies—the data gathering capacity is one of the things that we are working on in increasing in order to address some of these types of issues.

Maybe I will call on Mr. Peterlin to help me with any specific response as to what we are doing in that regard.

Ms. HOOLEY. Okay. That would be helpful.

Mr. PETERLIN. I can't honestly address the salmon issue itself, but what I can say is that you're talking data. And the data that is good for the salmon is the same data that is good for all the other aspects of USDA. I think NOAA is very well in tuned to the data that is necessary. And we are working as very strong partners with NOAA, meeting routinely. I would say that the data will not be lost and it will be available, but we do need the research that Dr. Gonzalez was talking about to check on the specifics of things like salmon, et cetera.

Chairman CALVERT. Would the gentleman identify himself for the record?

Mr. PETERLIN. I am Albert Peterlin, and I am the Chief Meteorologist for USDA.

Chairman CALVERT. Thank you.

Ms. Hooley, are you——

Ms. HOOLEY. Yes. Thank you.

Chairman CALVERT. Mr. Brown?

INSURANCE PROGRAMS

Mr. BROWN of California. Thank you, Mr. Chairman.

I wanted to follow up on a little bit of the line of questioning of Mr. Roemer about how to reduce the impact of the annual disaster supplementals through the use of insurance. I must confess that we have had pretty uniform agreement that an insurance program would be helpful, but we have had great difficulty in mounting insurance programs, particularly for earthquakes, which would mitigate the largest cost for the State of California.

The reasons for this are complex, and I am not the insurance expert. But I have been involved with earthquake matters for quite a few years. One of my desires has always been to reduce the burden on the federal taxpayers through the development of reasonable earthquake insurance with a strong mitigation component.

Now the problem that seems to exist is that the insurance industry likes the idea of more insurance, but they like it for the normal reasons. If it was required, it would produce a large increase in their volume and presumably they would manage to make a little money off of this. And if they had the strong backup from the Federal Government through secondary reinsurance or something of that sort, they wouldn't go broke in the event of a major disaster. But they don't care a great deal about mitigation, per se. In other words, tying insurance rates to improving the location of your structure, or the way your build your structure, or something of that sort.

And recognizing that none of you are insurance experts—particularly, Mr. Armstrong, you are not an insurance expert, are you? But you do have an insurance department or unit in FEMA that should be very much interested in this. We have been trying for several years to get an insurance bill through for earthquakes and other disasters—they ought to be really combined—and not with very much success. The only way we are going to break this log jam is if it gets high-level attention.

My question will be fairly simple. Do you feel that the time might be right for FEMA and other federal agencies that might be interested—this would include the Department of Agriculture because crop insurance has had the same problem. It is hard to devise a program that the farmers will buy as long as they know they are going to get disaster relief. So they don't buy it.

Do you think the time is right, as part of this large scale awareness, that we do need to reduce the size and level of participation of government in a lot of activities that can be handled by the private sector, to move ahead with something significant in this area?

Mr. ARMSTRONG. Well, you were right when you said that I am not an insurance expert.

Mr. BROWN of California. Yes, but this is a high level policy issue. And you are very good at high level policy issues.

[Laughter.]

Mr. ARMSTRONG. I think first that it would be well for the Congress to look at some of the issues I think you were alluding to in your remarks, which is the public receptivity to purchasing insurance, the public receptivity to even thinking about mitigation. There is such a short window of opportunity, it seems to me, after a disaster or right before one strikes where people are really paying attention, really feel the threat, and then maybe they will act.

In our society, it seems that having fire insurance as part of a homeowner's insurance plan is a natural activity in terms of what is purchased. But when you go into flood or earthquake or other areas there is a resistance on the part of the public. I think that making something palatable and also getting the insurance community to promote this is an issue as well.

I am from Colorado and we recently had some flooding out there. One of the follow-up stories out there was the confusion among potential policy holders to what they were hearing from insurance agents. I found the same thing in North Dakota when I was at the FEMA regional office in dealing with the Grand Forks floods. We need to do, I think, a better job of getting the insurance community on board as to what the current federal insurance programs could involve, and then have—I think you are raising an important issue. I don't know if the time is right or not.

Mr. BROWN of California. Yes, it's an issue which rises above the pay level of most people, as a matter of fact. You're not going to get public acceptance until you reach the point, as we have already in fire insurance, where you can't get a mortgage unless you have fire insurance.

Mr. ARMSTRONG. Right.

Mr. BROWN of California. We could do the same thing for earthquakes or floods or some of the other things we pay for by disasters, but it would require a national program in which everyone participates, and the burden would be shared based upon the risk of the particular individual or region where the insurance was applicable. You cannot get that without a national effort in which the Federal Government, the States, the insurance industry, and the other concerned people are all participating.

This is a question that arises in my mind: Are we ready to do this as a way of enhancing the efficiency of government and shedding responsibilities that ought to be carried by the private sector? You and Mr. Wheeler are probably two of the most important actors that could move in this direction, if you would feel highly motivated to do so.

Mr. ARMSTRONG. Well, certainly your interest is motivation enough for me, sir, and we will be glad to make sure we follow up and have more conversations on it.

Mr. BROWN of California. I wish Mr. Wheeler would say the same thing.

Mr. WHEELER. I will. And let me say that it shouldn't surprise you to know that California is already examining this question in the context of our experience with the floods.

I mentioned earlier that I served as the Chairman of the Governor's Flood Emergency Action Team. One of our recommendations was that we take a look at this whole question of flood plain management, and to make sure that local land use decisions—as we have already heard discussed—are consistent with the burden later imposed by those decisions upon state and Federal Government.

Mr. BROWN of California. Correct.

Mr. WHEELER. We are due to make some reports—including the implications for insurance of all of this—to the Governor in March. Let's hope that we're not sidetracked by our response to El Niño. But with respect to flooding, at least, we begin to understand the connection and we are exploring those opportunities.

LAND USE

Mr. BROWN of California. It has been pointed out that we are in a delicate area here where we are thinking of monkeying around with local land use decisions. That will immediately initiate a response that the Federal Government is becoming too intrusive and it won't sell. It will sell if there is a strong effort on the part of local governments in cooperation with state governments to develop some kind of standards as to how this would be done. Then the Federal Government becomes a partner invited in to make sure we get uniformity throughout the country.

This is an exciting opportunity to really make some progress that will help the taxpayers in the long run.

EARTHQUAKE INSURANCE IN CALIFORNIA

Mr. WHEELER. Mr. Brown, one more point about California's experience on earthquakes.

You know that the private industry has largely withdrawn from that market as a consequence of its exposure to risk in California. The legislature within the year has enacted a quasi-public authority for the purpose of making earthquake insurance more readily available. Because it is not part of my portfolio out there, I am not sure how that is progressing, but I would suggest that the real need, then, is for interaction between that state authority and a federal reinsurance scheme to assure that the State is not exposed to undue risk in that.

RE-INSURANCE

Mr. BROWN of California. Instead of setting up another quasi-government agency, the insurance industry itself, through a reinsurance program—in which maybe the Federal Government could be a part of the reinsurance—but I hope they could do it without setting up another agency—plays an role in it. And then the private insurance industry would carry the ball from there on.

Chairman CALVERT. Thank you, Mr. Chairman.

And as you know, our former colleague, Norm Mineta, had a bill that addressed that very issue on the secondary insurance market and reinsurance market. Then the late Bill Emerson took up the ball from there. And I think now Mr. Weldon is leading the effort for that disaster insurance legislation, which hopefully we will address later.

Mr. BROWN of California. We can encourage that, can't we?

Chairman CALVERT. Yes, I think so.

I have been very interested in this. I know that Mr. Hall just joined us. I know he has a question, so Mr. Hall?

Mr. HALL. I'll be brief because I have not heard all the testimony. I have read most of it or had it read to me. I don't know what questions have been asked. I would like unanimous consent to submit questions for the record.

Chairman CALVERT. Without objection, so ordered.

Mr. HALL. But I do have a question or two to Mr. Armstrong. I first congratulate you on being confirmed. I think that happened about 3 months ago, did it not?

Mr. ARMSTRONG. In June, yes, sir. Thank you.

REDUCING VULNERABILITY TO FLOODS IN TEXAS

Mr. HALL. And congratulate you for your association with Martha Braddock, who is my former FEMA advisor. She still advises me, but she is just not on our payroll.

You know the country—I don't know if you men—surely you have, and women, have heard it. But you hear murmurs about people being wary of rebuilding things that are flooded up and down the Mississippi River with tax money. They think they should have insurance coverage for that. I don't totally disagree with that. They are wary about rebuilding Los Angeles and San Francisco from earthquakes and the tornadoes and hurricanes that hit the east coast that are not common to some of the interior States. Texas is tornado and flood prone and they are wary of rebuilding some of our things.

I am not sure it is true, but someone told me that when Jarrell was blown away just the last 6 or 8 or 10 months, many people killed there, a little town totally destroyed—that they didn't get any federal support from that. You might check on that. That was told to me. I am not sure if that is true or not. If so, there may be a technical reason why they didn't.

But I guess the question I am asking is whether or not we really need legislation. You need flexibility, the flood mitigation assistance program, and other programs similar that are supportive there. They may need to be made more flexible and you may have covered that. But we had problems at Texoma with flooding. The Corps of Engineers handled that.

From time to time they would just flood the rice growers and everybody below. And from time to time they would flood out all the restaurants and commercial establishments above. I had both sides mad because the Corps of Engineers had rather specific instructions on when to start letting water out.

And some time ago, Wes Watkins and I—who is back in the Congress now, but he was a Member then from across the river in Oklahoma—we passed an act that said the Corps of Engineers could use their common sense. And we have not had that much problem since that time. They weren't given specific Congressional directions as to what level it had to be before they could start letting water out.

Maybe you might need something like that because Texas is flood prone. I just wonder what you recommend to be done to reduce the vulnerability to flooding. I know more notice and more equipment and things like that, but that is a question you can knock out the park. And I fell kindly to you and I want to submit it to you.

[Laughter.]

Mr. ARMSTRONG. Well, I think I answered a question from Congressman Roemer earlier. I don't know if you were in the room.

Mr. HALL. Well, he's going to ask me to explain it to him when I see him, so why don't you tell me again.

[Laughter.]

Mr. ARMSTRONG. He said I answered pretty effectively.

Mr. HALL. Seriously, he is a great Member.

Mr. ARMSTRONG. At the risk of repeating my answer to the Ranking Minority Member, mitigation programs are being implemented in Texas as well as across the country, and in Texas, through our regional office located in Denton. And I was down there for the Houston flood several years ago working on some community outreach programs. It was very interesting to watch the culture, at least at that time, where the joke was that people were carrying sheet rock back into their homes even as the floods were receding.

Getting the public to understand that after an event occurs there is a right way and wrong way to rebuild, so that when it happens again—and as you say, in Texas they know it is going to flood again. So in a sense, they are constantly on notice there. It is important that the partnership I have talked about and the other gentlemen here have talked about today is really pursued in terms of working with State and local governments and the private sector. We have to work hard with public information on how to rebuild and where to rebuild. And also how to learn to live with our environment. That certain things are going to happen. That is part of life, in a sense.

We do have some good success stories along the Mississippi River. We have moved entire communities or sectors of communities up to higher ground. And when those floods have come again—as they will—no damage occurred. In the case of Jarrell that you referred to earlier, to the degree that insurance covered the damages, that is accurate. There was no need for federal assistance because insurance covered the cost. And that is a benefit of having insurance. It reduces the impact on the Federal Treasury. But to the degree that humanitarian efforts were needed and to the degree that we could coordinate with the American Red Cross and with our state and local partners, we were there to assist. And we will be there. And we will always be there in a humanitarian sense.

But we do need to get people to take the initiative upon themselves, do some things with our federal programs so there is not an over-reliance on the Federal Government always being there for them. One of the examples is that we have modified our insurance program so that after an incident has occurred to an uninsured property, they are required to then purchase insurance so that if the incident reoccurs that they don't get the same amount of federal assistance. If they don't purchase insurance after the warning, so to speak, then their assistance is greatly reduced.

FIRE INSURANCE

Mr. HALL. That seems fair enough.

What about—from Mr. Wheeler—these poor people where they—they couldn't be poor people and have the homes that I see that are ravaged by fire in those divides—it seems like it is an annual matter that they lose their homes to fire there.

Mr. WHEELER. As has already been indicated, those people virtually always have fire insurance. One might argue that the insurance in that case is an incentive to rebuild, unless local government restricts the ways in which rebuilding occurs. I am told—although I am not authoritative on this point—that after the Oakland fires of the early 1990's, rebuilding occurred because local government allowed it to occur in ways that are not defensible, not fire safe.

We know a lot about how to plan so that we have access to structures that might catch fire, how to keep brush from around the foundations of buildings that would otherwise cause premature ignition. All of that is within our ability to disseminate to people and to motivate. But so long as there are counter-incentives—including the ready availability of insurance and decisions by local government which forestall their involvement in these issues—the burden falls back on state and Federal Government. We have a disconnect—a serious disconnect—between local actions and the burdens that are then borne by the state and Federal Government in response to those actions.

We need to build those bridges so that there is a better understanding of a quid pro quo, if you will, between those two.

Mr. HALL. Will Rogers said we was all ignorant, only on different things. And we all have problems, but they are just different—fire, flood, tornadoes, hurricanes, whatever.

I thank the Chairman for this hearing and I thank you gentlemen for giving your time.

I yield back whatever time I have.

Chairman CALVERT. Thank you, Mr. Hall.

And I want to thank this session for helping us understand the preparation that is taking place. As our abilities increase to more accurately predict future weather, our ability to prepare should also improve. We know that preparedness saves money. We all agree about that. And that is why I think we all support investment in long-term weather prediction so we can better prepare ourselves for the future.

Again, I thank you for coming today. This Committee is adjourned.

[Whereupon, at 12:12 p.m., the hearing was adjourned.]

[The following material was received for the record:]

HEARING OF THE SUBCOMMITTEE ON ENERGY AND ENVIRONMENT
COMMITTEE ON SCIENCE
U.S. HOUSE OF REPRESENTATIVES

on

Preparing for El Niño

Thursday, September 11, 1997

Post-Hearing Questions Submitted to

Dr. J. Michael Hall
Director, Office of Global Programs
National Oceanic and Atmospheric Administration

Limitations of ENSO Forecasts

Question 1. What are some of the limitations of current ENSO forecasts?

Answer: Some of the major shortcomings that are common to most of the forecast models are:

(1) Prediction of the magnitude and location of the sea surface temperature changes. Research shows that the accuracy of the atmospheric forecasts, for example, the likelihood of rainfall in California, depends on being able to predict the magnitude and location of sea surface temperature changes in the Tropical Pacific with reasonable accuracy.

(2) Accurate prediction of the timing of the onset and demise of the events. While the prediction of the general occurrence of an event is important, the accurate timing of the onset and demise is even more important. For example, the impact in the southwest United States of the 1982/83 El Niño occurred during late winter and spring. If the 1997 event should decay sooner than currently forecast, the forecast impacts would be less.

(3) Predictions of regional rainfall and temperature changes based on numerical models. Although early numerical climate models show features of the rainfall response over the United States, more research is required to predict specific regional changes in rainfall and temperature.

(4) Reliable Regional Forecasts. Ultimately, to accurately predict impacts such as flooding that would be useful for water resources managers, the regional deviations from normal rainfall and temperature have to be accurately predicted. We are a long way from being able to do this.

(5) Summer Climate Forecasts. The United States wintertime climate responses to ENSO events are reasonably well predicted. The nature of the summertime response and the ability to predict this are still very much a research subject.

Protocols for Issuing ENSO Forecasts

Question 2. The National Weather Service has developed protocols for issuing forecasts and warnings. What, if any, are the protocols for issuing an El Niño forecast?

Answer: The Climate Prediction Center (CPC) of the National Weather Service (NWS) has official protocols for issuing El Niño and seasonal and monthly rainfall and temperature forecasts for the United States. Usually on the fifteenth of the month, CPC issues an ENSO advisory as well as a monthly climate forecast for the next 13 months, which are posted on their web site.

ENSO advisory is posted at:

> http://nic.fb4.noaa.gov:80/products/analysis_monitoring/enso_advisory/index.html.

The monthly climate forecast for the next thirteen months is posted at:

> http://nic.fb4.noaa.gov:80/products/analysis_monitoring/ensostuff/index.html#us.f orecast.

In the past, the CPC forecasts were not integrated into the routine operational suite of products in the NWS regional offices. The size of this current event has initiated a host of activities in the NWS that will enable this to happen. The official products that are currently certified are based on statistical techniques that rely on observations. Within the next one to two years, these forecasts products will be replaced with dynamically produced forecasts based on our understanding of the physics of ENSO events. Additional work is required before most of the United States statistical inputs, some of which are not based on El Niño, can be replaced.

Improved Knowledge of ENSO

Question 3. Much of what we know about ENSO has been known for many years. In the intervening years, how has our knowledge improved?

Answer: Peruvians have known about El Niño for many years. In the 1960s scientists hypothesized a relationship between El Niño and the Southern Oscillation. Only after the initiation of an intensive international research program in 1983 did our knowledge improve in three critical areas:

(1) <u>Observations</u>: Our ability to better observe and thus describe the magnitude of ENSO events are due to the establishment of the ENSO Observing System in the tropical Pacific in the mid 1980s to early 1990s. (See Question 6 for a more detailed response to the importance of the observing system). For example, in 1982 without an observing system, we did not know that the century's largest ENSO was occurring until it was fully upon us. This year, however, due to the observing system, we were able to describe the evolution of the present ENSO in real time as it occurred.

(2) <u>Dynamical models</u>: In 1982, no dynamical models capable of simulating ENSOs existed. Over the past fifteen years, NOAA and their academic partners have developed intermediate and complex dynamical ENSO models allowing us to finally understand those physical processes that control ENSO.

(3) <u>Prediction</u>: Finally, in the early nineties with the advent of complex dynamical climate models at the National Centers for Environmental Prediction, it was possible to predict this year's ENSO several seasons in advance and ENSO-related climate impacts over the United States and the other parts of the world.

Ocean Observations vs. Ocean Modeling

Question 4. What are the respective roles of ocean observations and ocean modeling in ENSO research?

Answer: Our ability to predict ENSOs lies in our ability to model the ocean, which is the driver of the coupled ocean-atmosphere system. Thus, if we are to predict ENSOs accurately, we must be able to model the ocean as well as the atmosphere. Ocean observations are important, not only because they allow us to better understand, and thus better model the processes that control the evolution of El Niños, but also because they are a basic requirement for predictive models. The more accurate the ocean data, the more useful the predictive models.

ENSO Research Priorities

Question 5. What are the research priorities for NOAA's seasonal and interannual forecast program?

Answer: NOAA's research priorities for its seasonal to interannual climate program are:

(1) <u>Understanding Global Climate Fluctuations and Extreme Events</u>: NOAA will improve current predictions of global phenomena such as El Niño, which is often associated with droughts and floods around the world, including the U.S. This information is important for various reasons. For example, El Niño events are associated with decline of coho salmon on the West Coast. According to a NOAA funded research project, the economic value of improved El Niño forecasting to the coho fishery varies from $250,000 to $900,000 a year, depending on the accuracy of the forecast.

Critical to the success of the research is an ocean observing system. As the U. S. contribution to the Global Ocean Observing System (GOOS) and Global Climate Observing Systems (GCOS), NOAA will expand its ocean observing system beyond the tropical Pacific to other ocean basins, especially, the Atlantic and north Pacific.

NOAA will also increase efforts to translate understanding of natural climate variability on global scales into experimental regional forecasts and useable information on climate variability. NOAA will continue funding the International Research Institute (IRI) for seasonal to interannual climate prediction.

(2)   <u>Predicting U.S. Seasonal Climate Variability</u>:   This research is   important for prediction and mitigation of natural hazards such as the recent northwest floods, and drought in the southwestern U.S., and for planning purposes by agriculture, fisheries, and water and energy managers.

NOAA will advance our understanding of (1) how sea surface temperatures in the Pacific affect precipitation over the continental U.S. based on how natural climate variability is influenced by oceanic and atmospheric conditions over the Atlantic and Pacific Oceans, and (2) the hydrological cycle over the U.S.   NOAA will also accelerate the transfer of global scale predictions of El Niño to U.S. seasonal predictions of precipitation, runoff, soil moisture, and temperature.

In addition, NOAA will develop experimental regional climate forecasts and assess users' climate information needs and distribution networks. Currently, a NOAA-funded Columbia River watershed study provides seasonal climate information for use in water management, agriculture, and fisheries.

(3) <u>Explaining Climate Variability</u>: It is equally important to understand climate variability on longer time scales and the relationships to other time scales as well as regional climate variability. NOAA will focus on the oceans' role in climate and on understanding and documenting the natural variability of climate on decadenal and longer time scales. Two regional phenomena, the North Atlantic Oscillation (NAO) and Tropical Atlantic sea surface temperature variability, may hold the key to interannual to decadal climate variability over the U.S and Europe. NOAA's research will continue to make exciting new discoveries about the impact, origin and maintenance of long-term, multi-year climate anomalies in the Atlantic Ocean. This research may ultimately lead to predictive skill of climate variability and climate change over the Atlantic and surrounding continents.

TOGA Array

Question 6. What role does the TOGA array in the Pacific Ocean play in this research, and how is it critical to the success of the program?

Answer: The TOGA array, now called the ENSO array, is defined as *in situ* components of the TOGA Observing System, comprised of moorings, drifters, VOS/XBT and tide gauges. This tropical Pacific Ocean observing system is the glue that binds NOAA's ENSO research and forecasting efforts together. The oceanic and atmospheric data from this array connect the remotely sensed measurements from satellite to what is happening at and below the sea surface. Measurements from the array further connect the real ocean-atmosphere system to the virtual ocean-atmosphere system of climate forecast models.

The data from this array provide early warning of impending ENSO events, of which this year is an example. In late 1996, the suite of prediction models used for ENSO forecasting gave mixed signals about the likelihood of and ENSO warm event (El Niño) developing in 1997. Some predicted cold conditions, some warm, and some neutral. Data from the ENSO observing system were unambiguous. They showed an El Niño event exploding on the scene in the spring of 1997, at least a season in advance of what would be considered the normal development of El Niños.

In retrospect, those models that made the most use of the data from the array were also the models that showed the best skill in forecasting the current warm event. Thus, the continued maintenance of the present observing system is critical to ongoing improvements in the forecast skill of present generation computer models. The gains made in ENSO forecasting over the past 10 years can translate into hundreds of millions (perhaps billions) of dollars in savings from mitigation of lost productivity, property and agricultural output for the American taxpayer during future ENSO events. The President's FY 98 budget recognizes the importance of the ENSO Observing System and the need for continued research and requests $4.9 million to transfer the TOGA Array from research to operations, freeing up much needed research funds.

The TOGA array is the observational backbone for all further efforts directed at understanding the physical processes responsible for the genesis, evolution and termination of warm and cold ENSO events. Satellites and models are not enough. We must collect real ocean data over the entire expanse of the tropical Pacific, and over a long enough time to understand the basin scale ocean-atmosphere interactions that operate on seasonal, year-to-year, and decadal time scales.

Question 7. The National Research Council recently noted that, while TOGA was in general a successful program, it did not complete one of its objectives, which was the study of the tropical oceans. Rather, TOGA concentrated on studies of the Pacific Ocean. Are studies of other tropical oceans now under way or in planning? If so, please provide details on this research.

Answer: The follow-on research program to TOGA, the Global Ocean-Atmosphere-Land System (GOALS) program, has undertaken a study on the influence of the tropical Atlantic and Indian Oceans on climate variability. For example, NOAA is providing logistics support and training to the Brazilians and French for the deployment of moored buoys as part of an international effort to expand the Tropical Atmosphere Ocean (TAO) array from the Pacific into the Atlantic. These buoys are critical for the study of the variability of surface and sub-surface temperatures and winds at the ocean's surface.

This data will support statistical and modeling studies to understand the influence of Atlantic sea-surface temperatures on rainfall in the Americas and the processes that drive sea-surface temperature changes. With additional resources requested in the President's FY 1998 budget, NOAA will deploy additional buoys to complete the pilot array of some 13 moored buoys in the tropical Atlantic.

Projects are also underway to examine Indian Ocean variations and their relationship to climate variability in Africa, Asia and Australia. As with the Atlantic, a limited number of data analysis and modeling studies are targeting sea surface temperature variability; what gives rise to it and how does it affect climate variations in the region?

<u>Ocean Buoys</u>

Question 8.   Are other data-gathering ocean buoy arrays being planned or deployed?

A qualified '"Ayes."   NOAA research funding for climate observing systems is committed to the ENSO array (formally the TOGA array). The ENSO array must be maintained for operational climate forecasting. We cannot add new observations at the expense of the ENSO array. (See Question 6 for the importance of the array.) We do have plans, however, to expand the observing system as soon as the ENSO array is transferred to operational status (FY 1998 Presidential budget request for $4.9 million would make the ENSO array operational).

Question 9.   If other data-gathering ocean buoy arrays are being planned or deployed:

Question 9.1. Where will they be deployed?

Answer: As soon as NOAA receives additional resources requested in the President's FY 98 budget, it will begin to implement a new research array, "PIRATA" for Pilot Research Array in the Tropical Atlantic, similar to the ENSO array.   The total Atlantic Ocean observing system will be a composite of drifting buoys, floats, tide gauges, and Volunteer Observing Ships, as well as PIRATA moorings, and will extend observations into the North Atlantic; because a major mode of climate variability is the North Atlantic Oscillation (NAO) (See Question 10.2) which requires observations beyond the tropics.

Currently, NOAA funds a project being lead by Scripps Institution of Oceanography to integrate observations, modeling, and process research for improved climate prediction. Beginning in 1998 the project will deploy significant observational resources to improve understanding of the variability of ENSO. Approximately $1.1 million worth of drifting buoys, floats, and Volunteer Observing Ship observations will be used to expand ENSO array observations 600 miles north and south of the equatorial belt in the eastern Pacific.

Question 9.2. Who will be funding them and how much funding will be required?

Answer: The Atlantic array will be an experiment to learn how to observe the Atlantic modes of variability that are necessary to improve climate prediction. As such, the array will be funded primarily by NOAA and augmented by the National Oceanographic Partnership Program (NOPP) and the Small Business Innovation Research Program (SBIR).   The Atlantic array is estimated to cost $3.4 million/year ($2.2 NOAA, $1.0 NOPP, $0.2 SBIR).

The Pacific expansion of ENSO array observations north and south will be funded entirely by NOAA ($1.1 million).

Question 9.3. If it is an international effort, what will be the U.S. share?

Answer: The interagency U.S. Global Change Research Program recognizes NOAA as the U.S. lead agency for climate observations.   Similar to the ENSO Observing system, the Pilot Research Array in the Tropical Atlantic (PIRATA) will be an international collaboration.   This time France, Brazil, and the United States will be the principal players.   Partners in northern Atlantic observations are Canada, France, Germany, Portugal, Spain, United Kingdom, Iceland, and the United States.   Observations funded by the National Science Foundation and National Aeronautical and Space Administration, in addition to NOAA, will also contribute to the total Atlantic system.   Funding shares are: NOAA 45%, other U.S. agencies 25%, international 30%.

In the Pacific, operation of the present ENSO array (TOGA-TAO array) is an international effort with NOAA providing about 85% of the support.   The remaining 15% comes from Japan and France.   In 1998 Japan will begin a 3-year project to expand the array westward into the Indian Ocean, and northward into the northwest Pacific; Japan will assume operational responsibility for the 20 western-most moorings, while the United States and France will continue to operate the remaining 56 central and eastern Pacific moorings (about 95% U.S. and 5% France).   NOAA will fund the extension of the ENSO array in the eastern Pacific with drifting buoys, floats, and Volunteer Observing Ships.

Question 9.4. What do we hope to learn from these other arrays?

Answer: The influence of the Atlantic on our climate is thought to be as important as ENSO. The scientific understanding of Atlantic modes of variability is at the stage where understanding of ENSO was 10 years ago--there is much to be learned. The observing system will help scientists investigate the predictability of Atlantic variability and how it shapes regional climate effects, particularly in the eastern United States as well as Europe, Africa, and South America; hence the international participation in the observing system. The observing system will also be designed to specifically measure the hurricane genesis area of the tropical Atlantic to improve prediction of storm intensity and land fall tracks of hurricanes impacting the U.S. East and Gulf coasts.

ENSO is only one mode of climate variability in the Pacific (See Question 10.1.). Even though we can now predict ENSO with some degree of skill, accurately forecasting the specific seasonal impacts on the various regions of the United States, especially Alaska and Pacific Northwest, depends on understanding the interaction of ENSO with the other modes of Pacific variability and understanding the decadal modulation of ENSO as well.   The expansion of ENSO array observations north and south of the equator in the eastern Pacific is designed to observe and better predict how the tropical and subtropical Pacific are changing over seasonal to interannual periods, and to observe the decadal variations of ENSO and other Pacific modes of climate variability.

Pacific Decadal Oscillation and North Atlantic Oscillation

Question 10. The following set of questions refers to the Pacific Decadal Oscillation (PDO) and the North Atlantic Oscillation (NAO).

Question 10.1. Please explain the Pacific Decadal Oscillation.

Answer: Although El Niño is probably the best understood and most widely-recognized climate phenomenon in the Pacific, scientists have recently published their first findings on an emerging area of great interest of which we still know very little, "Pacific Decadal Oscillation," or PDO. Simply put, over a large area of the Northern Pacific, sea surface temperatures increase and decrease in step with surface air pressure and on decadal time scales. The PDO index was predominantly positive between 1925-46, negative between 1947-76, and positive since 1977. This interdecadal, coherent pattern of the ocean-atmosphere system appears to be related to surface temperatures and precipitation variations in North America and affects salmon catches along the Pacific Northwest coast. It would appear that the current low salmon harvest in the Northwest maybe, at least in part, due to the PDO.

Question 10.2. How does this climate pattern affect global climate variability and how do PDO and NAO interact with each other and El Niño and La Niña events?

Answer: How the PDO effects the global climate patterns is currently an area of research. Precipitation variations across much of North America can be shown to be statistically related to the PDO index, although the exact mechanism causing these relationships is not understood. The predominance since the late 1970's of the warm (El Niño) phase of the El Niño-Southern Oscillation phenomenon, along with negative sea surface temperatures over the extratropical North Pacific, accounts for most of the recent anomalous warmth over much of Alaska and western Canada and the relative coolness over the central North Pacific At this point in time, researchers have not observed any relationships between the PDO and NAO.

Question 10.3. How far are we from predicting PDO events?

Answer: Given the nascent state of the research and no new funds available to pursue such research, it is difficult to say when we will be able to predict PDO events and their connections to climate variability over the Pacific Northwest, western Canada and Alaska.

Question 10.4.   Please explain some of the challenges posed in making predictions of these types of events meaningful to resource managers?

Answer: Understanding the nature of the large-scale variations of the PDO, and understanding the effects on agriculture, water resources, and natural resources have the potential to be useful to policy-makers and institutions. Simply understanding which phase of the PDO we are in has important societal consequences. For example, we need to better understand the relationship between PDO and salmon productivity.

Question 10.5. Please explain the North Atlantic Oscillation?

Answer: Early this century, meteorologists noticed that year-to-year fluctuations in wintertime air temperatures on both sides of Iceland were often out of phase with one another. When temperatures are below normal over Greenland they are above normal in Scandinavia, and vice versa. Simultaneously, fluctuations in temperatures, rainfall, and sea level pressure acting in concert were documented, reaching eastward to central Europe, southward to subtropical West Africa and westward to North America. This mode of climate variability was then given the name: the North Atlantic Oscillation.

This phenomenon has been defined in terms of the strength of the Icelandic low and Azores high: when sea level is below normal in Iceland, it is generally above normal in the Azores, and vice versa. As a result, the NAO index is defined as the pressure difference between these two locations; above normal pressure in Iceland is defined as positive, or high. And negative, or low, when the pressure is below normal. It is now firmly established that fluctuations of the NAO influence climate from North America to Siberia and from Greenland to the equator and perhaps beyond.

Question 10.6. How does this climate pattern affect global climate variability and how do NAO and PDO interact with each other and El Niño and La Niña events?

Answer: Most meteorologists agree that the NAO is a primary climatic factor orchestrating hemispheric-scale climatic fluctuations centered over the Atlantic. Moreover, possible causes and the dynamics underlying these changes may now be emerging. A primary factor in modulating the NAO on decadal time scales may be its interaction with the underlying ocean.

While meteorologists have been puzzling over trends in Atlantic sea level pressure, subtropical droughts and the severity of European winters, oceanographers have discovered fascinating changes in the physical properties of the Atlantic Ocean. Variability in the water-mass properties of the Atlantic Ocean is almost certainly orchestrated by differences in the strength and path of storms at extreme phases of the NAO. For example, oceanographers have noticed that the process of deep and intermediate water renewal displays multi-year fluctuations in intensity, which occur synchronously with fluctuations in the NAO. More importantly, however, the ocean may not merely respond passively to the overlying meteorology. The possibility exists that, through air-sea interaction, the ocean may be responsible for modulating the phase and intensity of the NAO on decadal time scales.

Moreover, it is also clear that the NAO index has strengthened during the last 30 years or so, rising from its low index state in the 1940's and 1950's to reach a historic maximum high in the early 1990's. The NAO may now be beginning to wane - the past two winters exhibited a negative NAO index.

The NAO exerts a dominant influence on wintertime temperature and precipitation across the North Atlantic basin and thus has major impacts on marine and terrestrial ecosystems. The changes in the mean circulation patterns over the North Atlantic are accompanied by pronounced shifts in the storm tracks and associated synoptic eddy activity which affect the transport and convergence of atmospheric moisture and can, therefore, be directly tied to changes in regional precipitation.

It has been recently shown that drier conditions during high NAO index winters occur over much of central and southern Europe and the Mediterranean, while wetter-than-normal conditions occur from Iceland through Scandinavia. This has been the case for much of the past two decades when the NAO index reached an historic high in the early 1990's. Over the Alps, for instance, snow depth and duration over the past several winters have been among the lowest recorded this century, causing economic hardships on those industries dependent on winter snowfall, and severe drought conditions throughout Spain and Portugal has affected olive harvests. In contrast, increases in wintertime precipitation over Scandinavia may be related to increases in the maritime glaciers of southwest Norway, one of the few regions of the globe where glaciers have not been retreating. In addition, Norway, with a surplus water in its hydroelectric reservoirs, sold surplus electricity to other European countries.

The past two winters have exhibited a negative NAO index, hence northern Europe is drier and southern Europe had above normal precipitation while Europe as a whole has below normal temperatures. Due to drier conditions, Norway is meeting just its national hydropower demands. 1996 was a severe winter throughout Europe with record cold temperatures and heavy snowfalls throughout southern Europe.

Signals in eastern North American precipitation and runoff also show a relation to the NAO, which could impact a variety of economic sectors, especially water resources.

The combination of El Niño and NAO patterns contributes to the "Cold Ocean - Warm Land" (COWL) pattern that is the dominant spatial pattern of interdecadal temperature variability, particularly in accounting for Northern Hemisphere warming since the 1970's. To date, researchers have not observed a correlation with PDO.

Question 10.7. How far are we from predicting NAO events?

Answer: We are in the very early stages of research into the underlying causes of the NAO phenomenon. In order to be able to predict the NAO, coupled model studies that seek to understand the relationship between the ocean and atmosphere need to be an integral part of any research thrust, not the least because they will ultimately provide the route to studies of potential predictability. A critical objective must be to determine the characteristics, mechanisms, and predictability of decadal and multi-decadal variability observed in the current generation of coupled ocean-atmosphere models. It is crucial that this modeled variability be assessed in the light of the observed variability. The coupling processes in the Atlantic are subtler than in equatorial regions, and so the challenge will be greater than the one faced with understanding and predicting ENSO.

Observational evidence depicts a fascinating and complex picture of a climatically active Atlantic basin, in which ocean and atmosphere interact on a wide range of time scales to affect climate variability over the adjacent continents. It is of major importance that we better understand the dynamics of these fluctuations, study their predictability and quantify their impact on weather and short time-scale climate variability.   An investment in this research has the potential for large economic benefits similar to what we are beginning to achieve with the advancements in our understanding and prediction of ENSO.

Question 10.8.   Please explain some of the challenges posed in making predictions of these types of events meaningful to resource managers.

Answer: The indices of variability in the Atlantic Ocean and atmosphere must be further refined in order to be meaningful to resource managers. Further, the spatial and temporal properties of recurring patterns must be characterized. There must be an emphasis on basin-wide analysis so that local studies can be placed in a large-scale context.

One approach to obtaining large-scale distributions of properties is the reanalysis of existing data to generate gridded fields for use in computer models. The use of proxy records should continue to be actively pursued. These records are invaluable for assessing the fluctuations of climate during the pre-industrial era.

The strategy for understanding and predicting the NAO has a monitoring and process study elements. Monitoring to describe what is happening; observations of key processes to improve our understanding of them and their representation in models. Many time-scales are involved in the NAO and the physics depends on time scale. Observational strategies for the longest time-scales have to be different from those directed at a few years. Important meteorological as well as oceanographic issues are: (1) representation of the boundary layer in numerical weather prediction (NWP) models, and (2) failure of the NWP models to produce realistic oceanic heat fluxes.

<u>GOALS and GEWEXP Programs</u>

Question 11. In your testimony, you touch on the Global Ocean-Atmosphere-Land System (GOALS), the Global Energy & Water Cycle Experiment (GEWEX), and the GEWEX Continental-Scale International Project (GCIP).

Question 11.1. Please explain the goals of these programs.

Answer: The scientific objectives of GOALS are to (1) understand global climate variability on seasonal-to-interannual time scales, (2) determine the spatial and temporal extent to which this variability is predictable, (3) develop the observational, theoretical and computational means to predict this variability, and (4) make enhanced climate predictions on seasonal-to-interannual time scales.

The Global Energy and Water Experiment (GEWEX) Continental-scale International Project (GCIP) is a large continental-scale hydrometeorological study being carried out in the Mississippi River Basin. GCIP's mission is: "to demonstrate skill in predicting changes in water resources on time scales up to seasonal, annual and interannual as an integral part of a climate prediction system." To achieve this mission by 2010, GCIP has combined observational and modeling studies in meteorology, hydrology and coupled land-atmosphere modeling in order to improve the representation of land processes in both climate and numerical weather prediction models. In particular, GCIP emphasizes coupled model development, diagnostic studies, data assimilation and hydrological modeling for water resource applications, and data collection and archival.

Question 11.2. How do the GOALS and GEWEX Programs support the development of seasonal to interannual forecasts?

Answer: These two programs are designed specifically to improve predictions of seasonal-to-interannual climate variability. Together they test a central hypothesis that variations in the upper ocean, soil moisture, sea ice and snow exert a significant influence on seasonal-to-interannual variations of atmospheric circulation and thus, the predictability of the circulation. Model improvements for forecasting will rely on the improved understanding and representation of how variations in these so-called boundary conditions force variations in the atmospheric circulation.

<u>Predicting the End of ENSO Episodes</u>

Question 12. It appears that we know quite a bit about how El Niños start but are less sure of how and when these events fade. What are some of the scientific questions that need to be addressed to improve our ability to predict the end of an El Niño (or a La Niña) episode.

Answer: Some current forecast models already predict the end of El Niño and La Niña events. Issues still exist of the exact timing of onset and decay (*i.e.*, the events grow and die faster than predicted). The ocean-atmosphere interaction that causes these events to happen is rather crudely approximated in all models. Continued progress in our predictive capabilities will require improvements in: (1) simulation of the fluxes between the ocean and the atmosphere, and interaction between clouds, radiation, convection, and circulation in the tropical atmosphere, (2) details of the upper ocean mixing and boundary layer processes, and (3) ocean and atmosphere models use of data assimilation techniques to provide the initialization of the forecast.

Much of the increase in skill in short range weather forecasts has come from improved use of data through data assimilation for forecast initialization; the same can be expected for ENSO forecast models. The field of data assimilation for ocean models is still in its infancy and non existent for the couple forecast model.

<u>**Regional Impacts of La Niña**</u>

13. Please explain the regional climate impacts of a strong La Niña event.

Answer: The focus of ENSO research has been on El Niño impacts so less is known about a strong La Niña. The best known connection is with hurricane formation in the Atlantic Ocean. Given the progress in our knowledge and understanding of El Niño, one can expect that greater attention will be paid to understanding and documenting the impacts of the next La Niña, which some models are predicting could be as early as next year. The following examples will provide some insight to the impacts one can expect from a strong La Niña.

Tropics:

(1) <u>Hurricanes</u>: 1995, a La Niña year, witnessed 19 Atlantic tropical storms, 11 of which attained hurricane status, making this the second (next to 1933) most active hurricane season since records began in 1871. Tropical cyclones, such as the recent spat of hurricanes hitting the west coast of Mexico and the devastating tropical cyclone, Iwa, that hit Hawaii during the last major El Niño in 1983 would be unlikely in a La Niña year.

(2) <u>Federated States of Micronesia</u>: During La Niña years, the Marshall Islands experience low rainfall whereby salinity creeps up and can make the drinking water less potable or even unpotable in extreme cases of drought.

(3) <u>Chile</u>: During a La Niña, Chilean precipitation is below normal.

(4) <u>Peru</u>: La Niña years are welcomed by fishermen for the anchoyvetta return from Chile but not by farmers because these years have frequently been marked by drought and crop failures.

(5) <u>Columbia</u>: Both phases of ENSO have a tremendous impact on the hydro-climatology of Colombia. Warm phase (El Niño) is associated with diminished rainfall and river flows in the central, western and northern Colombia, and conversely for the cold phase. One of the rainiest years on record was 1971 which was accompanied by only a moderate La Niña event.

(6) <u>Mexico</u>: La Niña events often bring drought to northern Mexico.

(7) <u>Australia</u>: In 1988 a major La Niña event led to severe flooding in Queensland and New South Wales. The 1996 La Niña event, albeit a weaker one than in 1988, was correlated to extensive rainfall throughout eastern Australia in 1996.

The United States:

(1) <u>Pacific Northwest</u>: NOAA research has shown that the contrasting wet winter of 1993-94 and 1996-97 and the dry winter of 1994-95 in the Pacific Northwest were related to the ENSO phenomenon. We know from historical studies that during a La Niña event, when the waters in the tropical eastern Pacific Ocean become colder than normal, the Pacific Northwest experiences slightly above normal precipitation. During 1995-1996, the tropical Pacific was in such a cold phase of El Niño and the northwest received above normal precipitation causing flooding. NOAA's experimental model-based forecasts from the Fall of 1996 showed that the Pacific Northwest would experience above-normal precipitation that winter. While these research results do not necessarily indicate if or when an extreme flooding event will occur, they may signal a higher probability for flooding and provide important advance information to water managers and emergency preparedness officials.

(2) <u>The Southwest</u>: During La Niña events the southwest experiences diminished streamflow and snow water content. At Salt River, La Niña years are characterized by decreases in precipitation during winter and spring.

(3) <u>Great Basin</u>: La Niña is associated with significant droughts in the Great Basin. The drought of 1989 that extended throughout most of western North America was associated with a La Niña event.

(4) <u>Great Plains</u>: Florida State University research indicate that La Niña events increase tornadic activity in the Ohio and Tennessee river valleys. Results further show that El Niño inhibits the chances of multiple tornado outbreaks, while La Niña facilitates large tornadic outbreaks.

(5) <u>Southeast</u>: During La Niña events, the southeast receives below normal precipitation, making forest fires more likely.

Past El Niño and La Niña Events

Question 14. In your testimony, you document the occurrence of El Niño events in the past. Please provide a list of past La Niña events.

Answer:

ENSO PHASES

Cold Phase [La Niña]	Warm Phase [El Niño]
1904	1902
1909	1905
1910	1911
1915	1914
1917	1918
1924	1923
1928	1925
1938	1930
1950	1932
1955	1939
1956	1941
1964	1951
1970	1953
1971	1957
1973	1965
1975	1969
1988	1972
1995	1976
	1982
	1986
	1991
	1994
	1997

Note: Cold and Warm phases generally last into the winter and spring of the following year (*i.e.*, 1982 denotes the 1982/83 El Niño). Source: National Oceanic and Atmospheric Administration, Climate Prediction Center.

Seasonal/Interannual Climate Variability and General Circulation Models

Question 15. How do changes in seasonal and interannual climate affect the long term climate record and what are the implications for General Circulation Models?

Answer: Natural systems contain much inherent variability. As we go from the time scale of weather events to short-term climate, to seasonal cycles, to annual cycles to decadal and longer cycles, the climate signal present in each of these must be identified, isolated and filtered out to analyze any one of these time domains. Three key components of understanding the relationships between different time scales are:

(1) <u>Noise versus Signal</u>: If one is trying to understand the long term (decades and longer) climate record, the higher frequency noise in all of these domains must be filtered. Cycles such as the El Niño-Southern Oscillation (ENSO) is simply one of those time domains. The greater the amount of variability in the high frequency domains, the longer the record must be to fully characterize and understand long term cycles and changes. However, when major fluctuations such as the 1996-97 El Niño are well attributed to particular causes or cycles, the increased variation is easier to filter as it is part of a recognizable system.

(2) <u>Change Over Time</u>: We have seen important patterns of changing ENSO strength and frequency over past decades, and this variability remains an important area for further research.

(3) <u>Consistent Data</u>: The more difficult problems in developing a good long term climate record have been problems with changes in instrumentation and ways of observing climate that have occurred over the years. However, new techniques are being used to tease a clean climate record out of these data. Additionally, improvements in our understanding of past climate changes over even longer periods (decades to millennia) through paleoclimatic research are helping in this process. NOAA has funded a program to reanalyze weather data that will make a major contribution to improvements in this area.

Such variability in the climate system must then be reproducible in GCM simulations. The best GCMs are now capable of developing ENSO time-scale climate variability. This is essential to our ability to understand how the climate system functions. While we now have a much greater understanding of El Niño than we did a decade ago, we still do not understand what causes El Niños to form or what controls their intensity. The GCMs are still not able to simulate the full observed variability seen in the ENSO system. As this may be strongly influenced by climate activity outside of the tropics, GCMs will be essential to answering this question. Therefore, we are well served by the development of GCMs that are capable of simulating El Niños. This will both improve our confidence in the GCMs as we are able to simulate the observed climate variability and will improve our ability to study how various parts of the global atmosphere-ocean system influence El Niños.

Question 16. Have researchers been successful in eliminating the background noise of natural variability from these GCMs?

Answer: On the contrary, this is not the problem. GCMs are simplifications of our ocean-atmosphere climate system that are used to simulate what we observe. Because of that, GCMs must be improved to enable us to simulate this background noise that is seen in the natural climate record. However, an advantage of GCMs is their utility to model the simpler long term cycles and changes in the climate system and use these to better help us identify long term signals in our climate record and teasing these out from the natural and human produced noise.

Impact of El Niño and La Niña Events on Global Temperatures

Question 17. What impact do El Niño and La Niña events have on global average temperature? Do they increase or decrease average temperatures worldwide or do they simply shift temperatures spatially while leaving the average unchanged?

Answer: El Niño events result in higher global average surface and tropospheric temperatures than would otherwise be expected, and the opposite occurs with La Niña events. The El Niño-Southern Oscillation (ENSO) phenomenon is a primary mode of global temperature variability on time scales of 2 to 5 years. Excess precipitation during these events in many parts of the world leads to the release of latent heat into the atmosphere as water vapor is condensed into clouds and precipitation. Moreover, the extensive area of anomolously high temperatures in the equatorial Pacific contribute to significantly higher global average surface and tropospheric temperatures, especially when compared to the cold phase or the "La Niña" phase of the ENSO. For these reasons it is important to understand the stability of ENSO and its interaction and feedbacks with agents of global climate change such as changes in greenhouse gas concentrations.

Impact of Seasonal/Interannual Climate Events on Temperature Record

Question 18. Concerning the climate record over the last century, how much of the change in average global temperature do you think can be attributed to seasonal and interannual events such as El Niño, La Niña, the Pacific Decadal Oscillation, and the North Atlantic Oscillation?

Answer: Firm quantitative estimates are difficult to obtain on how much of the increase in global temperatures in the past century can be attributed to El Niño and other seasonal to interannual climate events, although recent studies indicate that contributions from such events may be substantial.

(1) Research shows that nearly all of the cooling over the northwest Atlantic and warming over Europe and Eurasia since the mid-1970's results from changes in the North Atlantic Oscillation (NAO), a mode which is characterized by considerable natural decadal variability.

(2) Over the Pacific-North American sector, surface temperature changes are more closely linked to decadal variations in El Niño, or more generally to tropical Pacific sea surface temperature variability. In particular, the predominance since the late 1970's of the warm (El Niño) phase of the El Niño-Southern Oscillation phenomenon, along with negative sea surface temperatures over the extratropical North Pacific, accounts for most of the recent anomalous warmth over much of Alaska and western Canada and the relative coolness over the central North Pacific.

(3) The combination of El Niño and NAO patterns contributes to the "Cold Ocean - Warm Land" (COWL) pattern that is the dominant spatial pattern of interdecadal temperature variability, particularly in accounting for Northern Hemisphere warming since the 1970's. Simulations with general circulation models suggest that much of the recent portion of the global temperature record (from the 1970s onward) can be reproduced by atmospheric models forced only by observed ocean surface temperatures.

The above results do not preclude the possibility that anthropogenic effects have altered the magnitude and frequency of occurrence of naturally-occurring seasonal to interannual events so as to make associated increases in global average surface temperatures more likely. A diverse body of evidence summarized in IPPC (1995) also indicates that global temperature increases over the last century are unlikely to be due only to secular variations in climate events such as El Niño and the NAO.

A key component in clarifying interpretations of the global temperature increases in recent years is the mechanisms that have produced the predominantly warm sea surface temperatures during this period. Dr. Kevin Trenberth and Dr. Hoar of the University Consortium for Atmospheric Research have argued that the persistent abnormally warm tropical Pacific sea surface temperatures (SSTs) during the 1990s are unlikely to be due to natural climate variability alone, although this interpretation is a matter of current scientific debate.

What is becoming increasingly clear is that there is a complex interplay between longer-term climate variations and shorter-term climate events such as El Niño and the NAO such that, in many cases, separate consideration of the two time scales is artificial. In particular, to provide reliable estimates of potential long-term climate changes, climate models should also be able to replicate adequately important natural modes of short-term climate variability, such as El Niño and the NAO.

Research Breakthroughs in ENSO Forecasting

Question 19. Are any major breakthroughs envisioned in the near future with respect of ENSO forecasting?

Answer: Although much of the progress in this area, as has been the case for short range weather forecasting, will be incremental, especially in the improvement in models and data assimilation systems, the following areas in which rapid progress might be expected are:

(1) Improved climate models: Improvement in the prediction of U.S. (and global) rainfall and temperature through improved atmospheric climate models and the coupling of these to regional and water resource models.

(2) Better understanding of all phases of ENSO: The El Niño and La Niña represent extreme states for the ENSO oscillation. Intermediate, less extreme states also impact the global atmosphere and U.S. temperature and rainfall distribution. Forecasts to capture these intermediate phases need to be more precise because the signals are smaller, and in some regions, like the western Pacific, current forecast skill is small.

(3) Role of other tropical oceans: Forecasts for the other tropical oceans and the impact of their sea surface temperature variations probably are important, especially on longer time scales (*i.e.*, the decadal variation).

(4) Understanding and predicting decadal cycles: The El Niño cycle has strong decadal modulation. Understanding the reason for this and being able to forecast the modulation are obviously critical for water management issues.

(5) <u>Improved summertime climate predictions over the United States</u>:   We understand very little, and hence, have low forecast skills for predicting El Niño impacts during the northern summer.

(6) <u>Understanding the link between ENSO and tropical storms</u>: We need to better establish the link between ENSO and tropical cyclone and hurricane activity using dynamical models and then use this link for forecasts of these phenomena.

Improving Value of ENSO Forecasts

Question 20. As we improve our ability to predict El Niño events, what needs to be done to improve the value of the information provided to customers of these forecast products, *i.e.* how can these forecasts be made more useful?

Answer: An improved understanding is needed of how decision-makers in climate sensitive sectors perceive their sensitivity to climate (are they willing to use new information), what their current coping mechanisms are, and what the opportunities are for integrating climate forecast information into their existing decision-making frameworks. For these reasons, NOAA funds research in the social consequences of and responses to climate variability as well as an emerging effort in regional assessments of climate variability. These regional assessments are designed to provide an interface between the users and producers of climate information as well as improve the regional resolution of climate forecast information.

Conveying ENSO Forecasts

Question 21. How does NOAA convey its predictions of El Niños to federal, state, and local agencies?

Answer: NOAA has two primary mechanisms, the NWS Climate Prediction Center (CPC), which produces the official United States forecasts, and the International Research Institute (IRI), which produces El Niño forecasts and atmospheric forecasts for its international partners. IRI and CPC share forecast information for the benefit of both. The CPC has issued seasonal forecasts (using statistical tools) for several decades (although El Niño forecasts have only been issued for the past few years). Hence the procedures for disseminating forecasts to other agencies and the private sector is well established. Currently, the primary mode is through Internet straight from CPC. (See Question 2 for web site addresses).

The size of the current El Niño, which suggests that heavy rains and perhaps flooding and mudslides might be associated with it, has raised the issue of how best to provide this information to local and Federal emergency managers. El Niño in essence changes the number and intensity of storms a given region might expect. The main function of the NWS has been to predict these and the potential for flooding. The local NWS field offices play a major role in communicating this information to local, state agencies and the public. Using the existing NWS infrastructure, NWS has begun the process of education and formal transfer of ENSO forecast products to their Regional Forecast Offices so that they may be an informed source for this information.

Reducing the Costs of ENSO Events

Question 22. How is NOAA working with other federal agencies to improve the usefulness of its seasonal and interannual forecast products to reduce the costs associated with these events?

Answer: Before this current event, NOAA worked primarily with technical experts from agencies such as Agency for International Development (AID), Bureau of Reclamation (BOR), the US Geological Survey (USGS) and US Department of Agriculture (USDA) to introduce them to the potential utility of ENSO forecasts. NOAA had also established pilot applications projects with universities in the northwest, the southwest, and Hawaii to work with local and state agencies in the use of climate forecasts.

When the projections for the large magnitude of the current event were made early in the summer, NOAA accelerated its outreach effort to AID, Department of the Interior (DOI), USDA, Federal Emergency Management Agency (FEMA), and Department of Energy (DOE). However, because of the wide press coverage and actions by members of the legislative body, agencies such as FEMA, DOI, AID, Corps of Engineers (COE) and, to a lesser extent USDA, became fully engaged in the use of such forecasts to help mitigate the potential impacts of this event.

NOAA is working with FEMA to develop its mitigation outreach program through the use of educational material, recorded and video messages, briefings, and compilation of past El Niños/La Niñas and natural disasters. NOAA has made presentations to DOI at the Deputy Secretary level where plans are being made to video conference to other DOI offices, and to a joint DOI/COE/FEMA/USDA/States of California meeting to discuss flood plain mitigation activities. Additional contacts are being made with technical representative of USGS, COE, and the State of California.

NOAA will soon begin weekly distribution to Federal and state agencies and to the public of ENSO update products to indicate areas that are under risk during the current event. NOAA is preparing a set of information packages for each State that has historically been strongly impacted by El Niño. These will be distributed to members of Congress and the State offices. Additionally the information will be available on the web for the public.

Costs and Benefits of ENSO Research

Question 23. What are the costs of seasonal and interannual climate change research and what are the estimated benefits?

Answer: Approximately 60% of NOAA's climate and global change budget is spent on seasonal to interannual research. In FY 1997, the amount was approximately $36,000,000.

Improved climate predictions will enable resource managers in climate sensitive sectors such as agriculture, water management, transportation, forestry and energy supply to alter strategies and reduce economic vulnerability. In 1982/83, when we had no predictive capabilities, the cost to society worldwide from floods, hurricanes, droughts, and fires from a major El Niño is estimated at $8,110,000,000. Early analyses of the potential use of ENSO forecasting in various management decisions have demonstrated the significant economic value of improved forecasts for different sectors in the US. Studies have:

(1) Estimated that the annual value of perfect ENSO predictions to the United States agricultural sector alone is $323 million (Solow *et al.*, accepted <u>Climatic Change</u>).

(2) Demonstrated the value of ENSO forecasting for management of salmon fisheries in the Northwest to be as much as $19.4 million over their 50 year planning horizon.

(3) Shown the usefulness of ENSO forecasts for forest management and hydropower operations.

When economic impacts beyond agriculture, such as energy, water, property destruction, and human health are taken into account, the estimated benefits of the forecasts will certainly multiply. While these studies give us a glimpse of the potential usefulness of these forecasts, much more research is needed to provide a thorough understanding of how decisions could be made more efficiently through the use of climate forecasts. This type of research would provide valuable information to decision makers across a range of sectors.

<u>International Co-operation</u>

Question 24. Please explain the extent of international co-operation for ENSO and other related research?

Answer: The ENSO observing system, the Inter-American Institute for Global Change Research (IAI), the Pilot Research Moored Array in the Tropical Atlantic (PIRATA), and the Pacific Ocean Observation and Research Initiative (referred to by its Japanese acronym TYKKI) are examples of international co-operation for climate-related research. Numerous countries support the ENSO observing system, drifting buoys, volunteer ship observations, and sea level measurements and tide gauges. The TAO array of moored buoys are supported by United States, Japan, France and Korea to list a few countries. Data from the ENSO observing system is nearly real-time available at www.pmel.noaa.gov/toga-tao/realtime.html.

IAI is an intergovernmental organization dedicated to augmenting the scientific capacity of the Americas, and to providing information in a useful and timely manner to policy makers. The United States National Science Foundation hosted the IAI Secretariat from September 1994 until September 1996 when the IAI became fully operational at the IAI Directorate located in São José dos Campos, Brazil. The IAI current research themes include ENSO and Interannual Climate Variability .

PIRATA is being implemented through multi-national cooperation. The purpose of PIRATA is to study ocean-atmosphere interactions in the tropical Atlantic that are relevant to regional climate variability on seasonal, interannual and longer time scales. France, Brazil and the U.S. are all cooperating in the PIRATA project.

TYKKI is a bilateral program between Japan and the U.S. in support of Pacific ocean observations and research. TYKKI's objectives are to promote and maintain ocean observation and research in the Pacific and surrounding oceans through cooperative activities between Japan and the United States.

International Research Institute

Question 25: Who are the other participants in IRI?

Answer: The principal U.S. players together with NOAA are Columbia University's Lamont Doherty Earth Observatory (LDEO) and University of California in San Diego's Scripps Institution of Oceanography (SIO). Over 144 countries have expressed interest in the establishment of an international research institution. Some of the more involved international players include countries such as Japan, Australia, and Brazil.

Question 26. How is IRI funded and what is the U.S. share, both nominally and as a percentage of funding?

Answer: The goal is to have the U.S. contribute no more than 50% of the total, with the remainder coming from a core group of other countries and potentially private sector contributors. In FY 1996, NOAA placed $4 million on the table which was 64% of the funds contributed that year. NOAA anticipates that its contribution will rise to $6 to $7 million by 2000 which would constitute less that 50% of the total contributions.

QUESTIONS FROM REPRESENTATIVE RALPH M. HALL (TEXAS)

Question 1. A great deal has been written recently in the press about the coming El Niño, and the contention that it may be the worst of the century. As I understand it, this may mean differences in weather conditions in different regions of our country while the El Niño is in effect. I am also a member of the Commerce Committee. My state generates large amounts of electricity and produces great volumes of natural gas. What effects will this El Niño have on the power consumption of the United States over the winter months, or energy requirements for air conditioning next summer?

Answer: El Niño years usually produce milder winters for most of the United States resulting in consumer savings in fuel bills. La Niña years usually produce above normal summertime temperatures resulting in an increase in fuel consumption from increased use of air conditioners.

An analysis of the effects of El Niño events on oil and gas consumption in the United States demonstrated that 1986-87, 1991-92 and 1994-95 El Niño events coincided with declining earnings for eight large North American oil and gas companies (source M. Wilson, "Trade Winds," Hart's Energy Markets). Because of the linkages between ENSO events and the energy sector, further research is needed to determine the consequences on the industry and the potential for the oil and gas industries to make use of ENSO forecasts. Very little is currently known about how U.S. power consumption will be affected by this El Niño. Analyzing these linkages would certainly be an area of high priority for NOAA's program.

Question 2. What kinds of changes will be seen in power consumption in other parts of the world particularly the developed world?

Answer: Certain parts of the world are highly dependent on climate-sensitive power sources, such as hydroelectric power. Because of its global effects, El Niño will impact many of these regions, some more significantly than others. ENSO events have a strong impact on portions of South and Central America. Hydropower is a significant source for energy in many of the countries in this region. Hydro-electricity constitutes the majority of national power for the region encompassing Costa Rica, Panama, and Colombia. For the region as a whole, El Niño events result in drought conditions. Because of the high dependency in these regions on hydroelectric power, the costs of a severe drought would be significant. NOAA is planning to undertake studies to better understand the implications of El Niño events for power demand and supply in various regions of the world.

Question 3. Have you established a working relationship with the Electric Power Research Institute (EPRI) or other utility companies to share your information and your knowledge? Is there data or information from those sources that would be useful to you that you are not receiving now?

Answer: We have had initial communication with EPRI in their climate and global change modeling division and power futures and have been working with various utility companies around the United States on the potential use of climate forecasts. Given EPRI's interest in climate impacts and the potential effects of ENSO events on the utility sector, an ongoing relationship with EPRI would be very beneficial. Various representatives from regional power companies, such as Illinois Power Company and Seattle City Light, as well as the Federal Energy Regulatory Commission have participated in NOAA constituent outreach workshops. NOAA also has connections with TVA and Bonneville Power to determine how climate forecasts could be made more useful for their management and planning purposes.

NOAA plans to more actively pursue connections with the utilities and other industries in the near future, such as briefings for various division with the Department of Energy and research programs to determine the sensitivity of this sector to ENSO and how decisions in this sector could be altered with forecast information. Another component of this research effort would be pilot studies in collaboration with utilities companies to determine the optimal way of using climate forecasts for the benefit of those companies. These studies could demonstrate to the broader industry how they might gain from climate forecasts.

Question 4. As you refine the predictive capability for the regions of this country and the world, are there plans to institutionalize information exchange and predictions with utilities, oil companies, gas producers, and the like?

Answer: El Niño forecasts are updated weekly. They are available from the home page of the Climate Prediction Center (CPC) (http://nic.fb4.noaa.gov). The seasonal forecasts for the United States are updated monthly and are also available to the public via the Web. Currently, no plans exist to provide special information exchanges with any specific economic sector. In order that the private sector, however, can have input into the usefulness of the forecasts, NOAA will undertake special outreach efforts. Two years ago when NOAA introduced the current format for the operational seasonal forecasts, the CPC along with the Regional Climate Centers held seven regional workshops to explain the product to the public and the private sector. NOAA is considering doing something similar once the current El Niño is over.

HEARING OF THE SUBCOMMITTEE ON ENERGY AND ENVIRONMENT
COMMITTEE ON SCIENCE
U.S. HOUSE OF REPRESENTATIVES

on

Preparing for El Niño

Thursday, September 11, 1997

Post-Hearing Questions Submitted to

Dr. Tim Barnett
Research Marine Physicist
Scripps Institution of Oceanography

State of ENSO Research

Question 1. Your testimony provides a good summary of the state of the science regarding El Niño. Where do we go from here and how many years away are we from getting there?

Answer: We are just entering the steep part of the learning curve with regard to El Niño forecasting and predicting the global impacts of this type of event. The Nation requires an aggressive research program aimed squarely at improving the prediction methods we have only recently discovered. I 'guess' such a program ought to be entering its mature phase in perhaps 10 years.

Research Priorities

Question 2. What do you see as the research priorities necessary to move our understanding of seasonal to interannual climate change forward?

Answer: The main priority is a coordinated program of climate modeling and prediction. While we now have many efforts in this area, they are either poorly coordinated or marginal in terms of expected success/impact. An associated priority is to have the best, fastest computers available to the modeling effort.

Question 3. In your testimony, you recommend the development of a national program that would apply long-range forecasts to benefit various sectors of the U.S. economy. How would the research community respond to such a program?

Answer: I believe the research community would welcome a national applications program. Its immediate impact would be to help them justify the research they are doing in financially tight times.   Such a program would also likely increase funding for applications of prediction.   Ultimately, I think the researchers would be attracted by the intellectual challenges and payoff to Mankind that such a program would offer.

Improving Value of ENSO Forecasts

Question 4. What level of detail is necessary to make these forecasts more useful to emergency planners, water managers, farmers, and other affected parties?

Answer: The people producing forecasts need to work with the eventual users to define the products to be forecast, for each application will require some specialized information. In effect, the forecasts/modeling needs to be 'regionalized' to account for the special conditions and needs of local user groups. Of critical importance to any forecast is the REQUIREMENT on the forecaster to supply some type of confidence measure on the forecast, *e.g.* the probability that the forecast will be correct. These measures of uncertainty will be of crucial importance to potential user/decision makers.

Preparing for ENSO

Question 5. Please discuss some of the measures potentially-affected cities, such as Los Angeles and Long Beach, are taking to prepare for El Niño. What type of support is Scripps providing these efforts?

Answer: Other researchers at Scripps are involved in this effort. I am still gathering information from them and will respond at a later date.

Question 6. You state in your testimony that countries like Australia, Peru, and Brazil have demonstrated the value of seasonal climate forecasts. Why, in your view, has the U.S. been slow to take advantage of these forecasts, most of which are developed in the U.S. and what can be done to improve the situation?

Answer: I do not know why the U.S. has been so tardy in using forecast information in its own interests. I think a number of agency leaders, businesses, etc. simply do not know what is available.   This could be cured by a well-coordinated national applications program and educational out reach effort. In other cases, Congress might need to prod status quo seeking bureaucrats.

Hazard Mitigation

Question 7. What agency do you see as being the lead agency in the *hazard mitigation* research area?

Answer: I think it is critical that a number of agencies work together on the use of seasonal to interannual climate prediction for hazard mitigation. These agencies should include at least FEMA, NOAA, USDA, Army Corps of Engineers, and the Department of Interior. It is also important that these agencies work with the academic community to ensure that research advances are transferred as quickly as possible to affected agencies.

ENSO Impacts on California Current

Question 8. It has been suggested that El Niño will have an impact on the California Current. How will this affect the currents and what are the expected economic impacts?

Answer: It appears the California Current is reduced in strength during an El Niño. This results in a warmer environment along the West Coast. Associated with this is reduced upwelling, reduced primary productive, and major shifts in planktonic communities. All of these factors attack the base of the food chain in oceans. Fisheries biologists hardly have even a rudimentary understanding of how such matters will impact fisheries resources in the eastern Pacific. Basically, we do not know the impact that El Niño has on many of the species that live in the eastern Pacific, let alone estimating the economic impact.

It is within our technology to model El Niño climate impacts in the eastern Pacific Ocean and include in such models various biological sub-models to ascertain impacts. The fisheries management community is essentially at 'GO' on this issue, *i.e.* they have not embarked on the necessary development actions to take advantage of current levels of understanding in this area.

Impact of Seasonal/Interannual Climate Events on Temperature Record

Question 9. In a recent article in *Science* by Richard Kerr (Vol. 276, 16 May 1997, pp. 1040-1042), you suggest that current GCMs do a fairly good job of simulating seasonal and interannual variability. How much of the warming observed over the last century can be attributed to climate variability?

Answer: How much of the recent warming can be attributed to Mankind and how much to natural variability? The answer depends on whose model you believe. In short, there is little agreement on this partition within the scientific community. The most current detection studies suggest the Man-induced signal is just now coming out of the natural noise of the climate system....although such statements have many caveats attached to them and rightfully so.

UNIVERSITY OF CALIFORNIA, SAN DIEGO   UCSD

BERKELEY · DAVIS · IRVINE · LOS ANGELES · RIVERSIDE · SAN DIEGO · SAN FRANCISCO   SANTA BARBARA · SANTA CRUZ

SCRIPPS INSTITUTION OF OCEANOGRAPHY   9500 GILMAN DRIVE   DEPT. 0224
CLIMATE RESEARCH DIVISION   LA JOLLA, CALIFORNIA 92093-0224

Climate Research Division, 0224
Telephone: 619-534-3223
FAX: 619-534-8561
E-mail: tbarnett@ucsd.edu

October 17, 1997

Representative Ken Calvert
U.S. House of Representatives
Committee on Science
Suite 2320 Rayburn House Office Building
Washington, DC 20515-6301

Dear Mr. Calvert:

During my testimony in early September, I said I knew of no studies that showed global warming might impact El Niño characteristics. Since then, an extremely sophisticated global warming simulation by the folks at the Max Planck Institute (MPI) in Hamburg has been completed and partially analyzed. The model that performed this simulation is arguably the best in the world.

The preliminary results from the MPI simulation show that La Niña (cold events) will become more frequent and stronger as global warming sets in. As I testified, these cold events have a uniformly bad impact on the United States (drought, tornadoes, hurricanes). Please remember that these results are from a numerical climate model. It is a very good, but not perfect, model.

I wanted to update you on the above matter since I did not have the information when I testified. Please let me know if you have any questions.

Sincerely,

Tim P. Barnett

cc: Representative Brown
     Representative Rohrabacher
     K. Ritzman

TPB/pdo

**HEARING OF THE SUBCOMMITTEE ON ENERGY AND ENVIRONMENT
COMMITTEE ON SCIENCE
U.S. HOUSE OF REPRESENTATIVES**

on

Preparing for El Niño

Thursday, September 11, 1997

Post-Hearing Questions Submitted to

**Dr. Andrew Solow
Director, Marine Policy Center
<u>Woods Hole Oceanographic Institute</u>**

<u>**Perfect ENSO Prediction**</u>

Question 1. Your testimony refers to perfect ENSO prediction. Can "perfect prediction" be achieved and, if so, what are the components of a "perfect prediction" and how long are we away from achieving them?

Answer: It is important to emphasize that my testimony dealt with the prediction of ENSO phase and not with other, subtler aspects of ENSO. While truly perfect phase prediction is impossible, it is probably possible to achieve very high skill with a lead-time of 12 months. Incidentally, although I did not cover it in my testimony, the work on which my testimony was based considered imperfect ENSO prediction as well.

Two components are needed for ENSO prediction: observations and models. In broad terms, the observations provide initial conditions for model simulations on which the prediction is ultimately based. There is another approach to ENSO prediction that is based on observations alone and does not use a model. In my view, however, the model-based approach, which incorporates the known physics, is more promising, especially in the long-run. For this approach, improved prediction arises from both improved observations and improved models. It is difficult to predict the pace of such improvements, but the appropriate time-scale of significant improvements in this area is 5-10 years.

Unavoidable Losses Related to ENSO

Question 2. In your testimony, you suggest that even with perfect foresight it would not be possible to avoid agricultural losses associated with an ENSO event. Over time, as experience with ENSO forecasts grows, would you expect these unavoidable losses to stay level, decrease, or increase and why?

Answer: I would expect them to decrease. In broad terms, a loss is unavoidable if the technology needed to avoid it is not available or is more costly to use than the loss itself. Experience with successful ENSO prediction should give rise to a demand for technology that can exploit this prediction. This demand should result in the development of the necessary technology.

Question 3. What makes unavoidable losses unavoidable and, if we are to get the most return from climate research, how can they be reduced?

Answer: As I said in answer to the previous question, in rough terms, a loss is unavoidable if the technology to avoid it is missing or costly. ENSO prediction will generate a demand for new types of technology—types specifically designed for responding to long-range climate prediction. Public and private investments in research and development in this area would increase the value of ENSO prediction.

Benefits of El Niño

Question 4. You referred to the benefits certain areas of the country will reap because of El Niño events. Was the increased ability of businesses and governments to take better advantage of these benefits included in your calculations for the economic benefits of El Niño prediction?

Answer: Absolutely.

Improving Value of ENSO Forecasts

Question 5. Improving the science supporting ENSO forecasts will obviously improve the value of the forecasts. What in your view needs to be done to improve the use, or application, of these forecasts by federal, state, and local governments, and so increase their value in this way.

Answer: Although dissemination is important, I believe that there is a tendency to overstate its importance. In the U.S. economy, the greatest value for ENSO prediction will arise from its use by decisionmakers in the private sector. The private sector is not unsophisticated. Agricultural firms, insurance companies, and power-generating companies often hire their own meteorologists and are, therefore, in a good position to make use of ENSO predictions. Even small-scale farmers pay considerable attention to weather predictions. I am less familiar with decisionmakers in the public sector, but my impression that they, too, are in a good position to use improved predictions.

Uncertainty of ENSO Forecasts

Question 6. The uncertainty inherent in ENSO forecasts obviously plays a role in their value. How do people incorporate this uncertainty in their decisionmaking, how does this affect the value of ENSO forecasts, and what can be done to increase that value?

Answer: In our work, we assumed that decisionmakers maximize expected profit (or net benefit). This is the standard assumption about decisionmaking under uncertainty. For example, suppose that the prediction is that there is an 80 percent chance of an El Niño event and a 20 percent chance of no event. In calculating expected profit for a particular decision, the decisionmaker does not simply assume that an El Niño event will occur, but takes into account the 20 percent chance that one will not, in which case a strategy that simply assumes that an El Niño event will occur may do badly.

In general, the forecast has higher value the less uncertain it is. Thus, value can be increased by investing in improved prediction. However, there is a diminishing return to prediction skill, so that an improvement from low skill to moderate skill has more value than an improvement from moderate skill to high skill.

Potential Role of Commercial Sector in ENSO Forecasts

Question 7. Your testimony suggests that ENSO predictions comprise a good example of a public good and that it is a legitimate role of government to provide them. Do you see an eventual role for the commercial sector in providing specialized and localized ENSO predictions similar to the role the commercial sector plays in providing specialized and localized weather forecasts?

Answer: I do. First, the more specialized the prediction, the less "public" a good it is. Second, the less useful the prediction is to the wider public, the less need there is to keep it secret.

El Niño Impacts on Fishing Industry

Question 8. Your testimony refers to the impact of El Niño on the fishing industry. Please explain how ENSO affects the fishing industry in the U.S.

Answer: We conducted a study of the value of ENSO prediction to the coho salmon fishery in the Pacific Northwest. We found that natural mortality, particularly of juveniles, was higher during El Niño events, mainly as a consequence of reduced summer streamflow. Advanced knowledge of this could be used by the manager of this fishery in setting total allowable catch to maximize better the long-run value of the fishery. The specific management response to this information depends on the life-history and population dynamics of the particular fish stock, so that what is optimal for salmon is not necessarily optimal for other species.

El Niño Impacts on Major Industries

Question 9. What are some of the other major industries that could be affected by El Niño events?

Answer: In addition to agriculture and fisheries, I expect that the hydroelectric power-generating industry in the southeast and northwest could make use of ENSO prediction for flow and reservoir management. I would expect that ENSO prediction has value to water management generally. It is also likely that industries involved in heating and cooling (e.g., oil, natural gas, electric power, etc.) could use ENSO prediction for planning purposes.

El Niño Impacts on Fisheries Programs

Question 10. What are the implications of El Niño and La Niña events on programs to maintain a viable, sustainable fishing industry and are these programs adequately flexible to contend with such events?

Answer: ENSO affects the dynamics of certain fish stocks, like the Pacific salmon. In general, the more predictable fluctuations in a fish stock are, the better it can be managed. Having said that, the value of ENSO prediction to fisheries management must be a small fraction of the value to rationalizing management itself (e.g., by implementing individual tradable quotas or other market-based management approaches to fisheries management). If management is rationalized, then the value of ENSO prediction may actually increase in absolute terms, as the overall value of the resource increases.

Value of ENSO Predictions to Foreign Countries

Question 11. You note in your testimony, and we have heard from other witnesses, how valuable ENSO predictions are in other parts of the world. Indeed, you suggest that providing these predictions should be considered a form of foreign aid. What do you estimate is the value of these predictions to foreign countries?

Answer: To my knowledge, no serious attempt has been made to answer this question. We are currently trying to develop a project to answer this question for countries in Latin America and also for China. In very rough terms, I would guess that the value in Latin America is $1-5 billion per year. I arrive at that figure by considering, first, that the effects of ENSO are greater in Latin
America than in the U.S. and, second, that the economies in Latin America are smaller than that of the U.S. I cannot make a guess for China.

International Co-operation

Question 12.   What is the contribution of other countries to ENSO and other seasonal and interannual climate research?

Answer: The European nations have active research programs in climate prediction. Because the recognized effects of ENSO on climate in Europe are relatively minor (with the possible exception of Spain), this effort is not as closely tied to ENSO prediction as ours.   The Australians also have efforts underway that are closely tied to ENSO prediction.   Communication between U.S. groups and Australian groups is quite good.   I cannot say the same for European groups.   I understand that the Japanese are also mounting an effort in this direction, but I know nothing about it.   To speak baldly, I would say that the U.S. is clearly the leader in this area.   I attribute this in part to the relative freedom enjoyed by the U.S. research community in pursuit of the advancement of science.

**HEARING OF THE SUBCOMMITTEE ON ENERGY AND ENVIRONMENT
COMMITTEE ON SCIENCE
U.S. HOUSE OF REPRESENTATIVES**

on

Preparing for El Niño

Thursday, September 11, 1997

Post-Hearing Questions Submitted to

**Mr. Michael Armstrong
Associate Director, Office of Mitigation
<u>Federal Emergency Management Administration</u>**

<u>**Interagency Communication and Climate Forecasts**</u>

**Question 1. How does FEMA interact with NOAA and the research community
responsible for the development of these seasonal and interannual forecasts?**

Answer: FEMA has an close and continuing relationship with NOAA, including the
Weather Service and the National Hurricane Center, and others involved in weather
forecasting and climatic research. In fact, FEMA has access to nearly every piece of data
and information generated by NOAA, and both meets with and communicates with the
organizations involved with weather forecasting on a regular basis to discuss trends,
potential problem areas, and new informational requirements in the emergency
management community for weather-related data. FEMA also has a National Hurricane
Team that works with NOAA's National Hurricane Center (and collocates with them
during disasters), to maintain a strong relationship between our two agencies and ensure
that State and local officials have the best and most timely information about hurricane
formation, storm location, and both short- and long-term climatic trends.

In addition, FEMA works closely with NOAA on development of their information
dissemination systems to better ensure that available weather data and research reaches the
customers that most require that information, and that it meets their needs.

<u>Preparing for Climate Events</u>

Question 2.   As our ability to predict seasonal to interannual climate change improves, what are some of the things FEMA will be doing differently than it has in the past to prepare for severe climatic changes?

Answer: The primary difference from the past will be the Agency's emphasis on mitigation as a means of reducing potential disaster losses.   As weather and climate predictions improve, it will become easier and easier to forewarn communities about the real risks they face in the coming weeks, months, and years.   This will open the door for FEMA to work with State and local governments to help them identify mitigation opportunities, prioritize their efforts, and identify local, State, Federal, and private sector resources to reduce their risk in advance of the next disaster.

In fact, this is the primary emphasis of Director Witt's new initiative, *Project Impact*.   The Congress has given FEMA funding to encourage communities to develop partnerships, identify risk reduction activities in order to reduce the potential costs of future disasters, and leverage resources from the private sector, the States, and other Federal agencies to make their communities more disaster resistant.

We will never be able to eliminate all disaster damages, but we can reduce them and help our communities become safer, stronger, and smarter.   But while the opportunities are there for the taking, the cultural shift that will be necessary to make mitigation part of day-to-day community decisionmaking cannot be created on a national basis overnight.   For these reasons, improved climatic forecasts will also allow FEMA to help prepare the American people and their local communities for possible future disasters through educational campaigns, the provision of technical assistance, and by providing training opportunities.   A perfect example of FEMA's commitment to assisting in this regard was the recent El Niño Summit, held in California last month.   This Summit served as a valuable forum to raise the awareness of community leaders and citizens in the western United States who may experience devastating impacts from the El Niño over the winter months.   The results of this effort are already paying off with increased attention to mitigation and emergency preparedness.

<u>Improving Value of ENSO Forecasts</u>

Question 3.   Given the type of information available in current NOAA forecasts, what type of information and level of detail would FEMA like to see in future forecasts that would improve its responsiveness?

Answer: At this time, FEMA does not believe that any major changes are needed with regards to the type and level of detail for information provided by NOAA forecasts.   In addition, in those instances when FEMA required additional information or clarification about forecasts or climatic research, NOAA has demonstrated a true willingness to be flexible to meet any additional needs we have.

But like everyone else at the Federal, State and local levels, the primary need FEMA has at this time is for NOAA to continue their efforts to improve the accuracy of long-term forecasts so that there can be greater certainty of what problems may arise in the distantly in the future. This could significantly improve FEMA's ability to be responsive to State and local needs, and would permit more aggressive preparedness activities and mitigation outreach in advance where future disasters may occur.

Preparing for El Niño at FEMA

Question 4. It appears from your testimony that FEMA will primarily depend on existing programs and policies, with some modification, to deal with the upcoming El Niño. Please explain how, or if, FEMA has shifted resources to deal with this event.

Answer: FEMA has gone to great lengths to address the upcoming El Niño. For instance:

- Our Regional Offices in Denver, Colorado; San Francisco, California; and Bothell, Washington have been working diligently with their States and local communities to educate them about various types of mitigation techniques that can be applied to lessen the impact of El Niño.
- In October, Vice President Al Gore and FEMA Director Witt held an El Niño Summit in California in order to provide a forum for Federal, State, and local officials to share information and prepare for potential damages over the winter. This event was highly successful, not only in terms of raising public awareness, but also in terms of flood insurance sales (the number of policies sold in California, for example, grew by over 11 percent, and the volume of telephone calls we received from people interested in purchasing flood insurance has more than doubled).
- A significant marketing and media affairs effort has been initiated, with special emphasis on the States most likely to be impacted by El Niño, to encourage people to reduce their risks and purchase flood insurance. This includes advertising on television, radio, and in periodicals.
- National Flood Insurance Training Workshops for insurance agents and lenders are being held to increase their knowledge of the flood insurance program and help ensure the timely handling of flood insurance claims.
- FEMA is conducting training workshops in numerous locations on the west coast and southwest to teach local community officials proper floodplain management principles and practices.
- We are conducting Community Assistance Visits in over 100 communities to assist them in evaluating and strengthening their floodplain management programs.
- FEMA floodplain management experts have been contacting communities that are not yet participating in the National Flood Insurance Program in an effort to get them to adopt minimum floodplain management standards and participate in the program.

- FEMA has developed its "El Niño Loss Reduction Center" on the Agency's internet site, emphasizing preparedness and education about the El Niño phenomenon and potential impacts.
- The Agency developed El Niño information packets, which are available from the agency's toll-free publications warehouse number (1-800-480-2520).
- The Agency is reviewing and strengthening its staffing and deployment plans, and is working with States in an effort to prepare for the possibility of multiple concurrent events.
- We are actively working with at-risk States to ensure their Hazard Mitigation plans are current.
- FEMA staff working on disaster closeout activities are refocusing their activities in order to expedite the resolution of prior event flood-related claims. This is being done in order to speed the repair and mitigation of previously damaged facilities in advance of any future disaster.
- We have replenished the Agency's "go kits" and have pre-identified resources that may be necessary in initial response operations, should they be necessary.

While these are but examples, they demonstrate that FEMA is utilizing its existing resources and programs to address the threat of El Niño.

Flood Mitigation Assistance Program

Question 5. In your testimony you mentioned that under the National Flood Insurance Reform Act of 1994, the Flood Mitigation Assistance Program was created to provide Planning Grants to assist communities with updating flood mitigation plans and Project Grants to implement measures to reduce floods. Has FEMA altered, or does it plan to alter, its pattern of funding to take into account the El Niño prediction and to focus a greater portion of its assistance on areas expecting greater-than-normal rainfall because of El Niño?

Answer: FEMA is working with the States with the highest potential of being impacted by severe weather this winter to encourage them to target their FMAP funding to address those structures most at risk from the El Niño threat. At this time, however, FEMA is not planning to adjust the funding pattern for the Flood Mitigation Assistance Program (FMAP).

FEMA has already promulgated regulations which provide a formula for the distribution of FMAP funding to the States. Under this formula, each State receives a base amount ($110,000) of funding each year, with adjustments upward based on the number of flood insurance policies in force in that State (so that the people who are paying for the program benefit the most) and the number of repetitive loss structures (so that funding can be focused in those areas where the most high-risk structures are located). Under this formula, we feel the States are treated fairly based on the risks they face and the history of repetitive losses in their jurisdiction. This is consistent with the direction provided by the Congress in the passage of the National Flood Insurance Reform Act of 1994.

Question 6. In your view, does the Flood Mitigation Assistance Program (FMAP) and other similar programs need to be made more flexible to deal with the predicted effects resulting from seasonal to interannual climate change?

Answer: Programs such as the FMAP are already highly flexible, and can be used very effectively to address the predicted effects from such climatic changes. In fact, the funding provided to the States under FMAP is provided to address repetitive loss structures, without few other restrictions on where and how that funding should be utilized. This provides States with tremendous flexibility to meet their own priorities in reducing flood losses. As a consequence, they have ample opportunity to use FMAP funds to respond to State and local mitigation requirements in the face of seasonal and interannual climate change.

The same is true for FEMA's other pre-disaster mitigation activity, *Project Impact*. This initiative is designed to allow local communities to identify their own risks and set their own priorities on how to address them. To assist them in this process, FEMA provides seed funding, and works with the local community to leverage that in order to gain the support of and utilize resources from the private sector, as well as local, State and Federal government to address their needs. Right now, FEMA is already working with the localities in the San Francisco Bay area, as well as in Seattle, Washington, in order to assist them in becoming more disaster resistant communities. In addition, our FEMA Regional Offices will begin working with at least one community in each State during this fiscal year to encourage them to follow the *Project Impact* model.

Improving Program Flexibility

Question 7. How can flexibility be built into these programs to improve their responsiveness to seasonal and interannual events like El Niño?

Answer: As previously discussed, the existing programs are already very responsive, and little additional work needs to be done. In addition, we are constantly meeting with and seeking input from the States, other Federal agencies, the Congress, local governments, representatives from the private sector, and others to ensure our programs continue to be responsive to their needs and priorities for risk reduction as they change over time.

Interagency Communication and Climate Forecasts

Question 8. Your testimony includes examples of cooperation between NOAA and FEMA in a number of areas. Please explain in more detail the relationship between NOAA and FEMA concerning ENSO predictions. Are formal contacts routine or are they on an *ad hoc* basis?

Contacts with NOAA on long-term climatic trends and analyses, including ENSO predictions, are routine. FEMA and NOAA have worked over the past several years to develop a strong partnership wherein we share information and communicate constantly (to include the communications of ENSO data) over the phone, electronically, and in periodically scheduled meetings. FEMA utilizes the long-term prediction data provided by NOAA to monitor trends, develop response plans, contribute to modeling efforts, focus the provision of technical assistance to the States and local governments, and even help establish priorities for the Agency's education and outreach initiatives.

Insurance Industry

Question 9. In your view, how has the insurance industry responded to the improvements in the seasonal to interannual forecasts of severe climatic changes?

Answer: Unfortunately, this is a questions that FEMA is not able to answer. How the insurance industry utilizes such forecasts is considered proprietary information within insurance organizations.

Self-Insurance Fund

Question 10. It was suggested during the hearing that FEMA establish a self-insurance fund with all the standard practices appropriate for setting up and operating such a fund. Has FEMA examined the possibility of establishing such a fund and would such a fund be an appropriate vehicle for reducing the cost to the taxpayers of emergency relief efforts?

Answer: In a unique opportunity afforded by the coordination of activities under both the National Flood Insurance Program and the Disaster Relief Program, FEMA has been able to require that State and local jurisdictions have a minimum amount of insurance for flood hazards. In the event of a declared flood disaster, FEMA will subtract the amount of available flood insurance coverage from disaster recovery assistance in damaged public buildings. This approach may also work for hurricane, earthquake, wildfire, and other natural hazards, and we are exploring the implementation of incentives that could cause State and local jurisdictions to carry out some level of insurance coverage on public structures.

However, it is also crucial to note that these measures will not reduce the cost of disasters. The purchase of insurance will shift who pays for disaster losses, and when they pay, and this shift may be somewhat appropriate in that those at risk bear a larger share of the burden. But only the implementation of risk reduction measures, mitigation, will reduce the ultimate cost to taxpayers of disaster relief efforts.

HEARING OF THE SUBCOMMITTEE ON ENERGY AND ENVIRONMENT
COMMITTEE ON SCIENCE
U.S. HOUSE OF REPRESENTATIVES

on

Preparing for El Niño

Thursday, September 11, 1997

Post-Hearing Questions Submitted to

Dr. I. Miley Gonzalez
Under Secretary for Research, Education, and Economics
United States Department of Agriculture

<u>**Cooperative Extension Service**</u>

Question 1. How is USDA getting the word about the current El Niño to U.S. farmers, and what is the role of the Cooperative Extension Service?

Answer: The National Weather Service (NWS) Climate Prediction Center (CPC) issues periodic El Niño advisories. A staff of CPC meteorologists are assigned to the USDA/Department of Commerce Joint Agricultural Weather Facility (JAWF) and keep USDA analysts informed of the latest global weather conditions that may affect agriculture. Weekly briefings are offered to USDA administrators and staff to ensure that agencies are currently aware of the latest conditions that may impact their missions. In addition, NWS and USDA jointly publish the Weekly Weather and Crop Bulletin electronically and in hard copy format for public dissemination. Radio and TV interviews are also conducted routinely. The Cooperative Extension Service serves a vital role in the delivery of weather and climate information and management advice to local agricultural community.

<u>**Preparing for El Niño in Agricultural Sector**</u>

Question 2. What has been the general response of U.S. farmers to this information and to your knowledge have they altered their normal practices in any way in preparation?

Answer: Farmers get most of their weather and climate information from the NWS. USDA provides timely information on the impact of weather and climate on agriculture, generally with examples of analog or similar years for comparison. Accurate data and information give farmers an increased knowledge base to prepare for alternative strategies. The impact of El Niño is well documented in some regional production areas.

However, as Dr. Hall testified before this committee, U.S. weather is highly variable and in any pattern of weather over North America, El Niño is only part of the story. While growers might respond to weather and climate forecasts if they were timely and accurate, the 1997 El Niño event unfolded well into the spring and it would have been difficult for growers to respond. Finally, it should be noted that although much more research must be accomplished, anecdotal evidence suggests many of the effects of El Niño might be positive, that is, precipitation above normal levels.

Improving Value of ENSO Forecasts

Question 3. Given the type of data available in current NOAA forecasts, what type of information and level of detail would USDA like to see in future forecasts that would improve its responsiveness?

Answer: Accurate forecasts issued well in advance of occurrence would be of greatest benefit to agriculture. However, even perfect forecasts will not eliminate all risk for growers. In addition, forecasts of general trend, above or below precipitation and above or below normal temperature, while very useful, must be more specific with respect to timing and intensity before crop yield forecasters can make better use of them. Finally, USDA continues to need timely access to real-time global weather and climate observations along with precise quality control and a thorough diagnostic analysis of the actual data.

Dissemination of weather and climate information to the agricultural community is one of USDA's principal goals.

ENSO and Agriculture in Brazil

Question 4. What has USDA learned from the experience of farmers in northeastern Brazil, who used seasonal climate forecasts and modeling developed in the U.S. to make adjustments to normal framing practices to maximize yields under stressful environmental conditions?

Answer: USDA is keenly aware of the experience of farmers in tropical latitudes, such as northeastern Brazil, using seasonal climate forecasts. It is more difficult to apply the same approach to higher latitudes that do not experience alternating wet and dry seasons governed by the inter-tropical convergence zone that migrates with the seasonal position of the earth relative to the sun. In higher latitudes, i.e., the United States, the strong upper-level winds, commonly referred to as the "jet stream," provide the steering mechanism for storm systems that are also strongly influenced by regional sources of moisture and topography.

International Co-operation

Question 5. Has USDA been working co-operatively with other countries to exchange information and minimize the impacts of El Niño?

Answer: Yes, USDA is working closely with several countries. For example, through the U.S./South Africa Bi-National Commission, USDA is working with South Africa to improve its data collection, analysis, and dissemination procedures. South Africa is one of the countries most often affected by El Niño during its corn growing season.

In addition, several USDA research laboratories and experiment station scientists conduct research on El Niño, often in collaboration with other countries. USDA's Southwest Watershed Research Center (SWRC) will be working with Mexican officials to develop El Niño-Southern Oscillation (ENSO) linked parameters for stochastic daily weather models. These models serve two purposes: to produce daily weather variables for inputs to agricultural and natural resource models; and to identify statistical linkage of ENSO to local and regional weather.

Finally, USDA provides frequent briefings to other countries and cooperates with many countries through the United Nation's World Meteorological Organization on agricultural weather applications.

Value to ENSO Forecasts to USDA

Question 6. Granted that the impacts of El Niño diminish in temperate regions, do you think that USDA been slow to recognize the potential value of seasonal to inter-annual climate forecasts?

Answer: No. USDA has carefully scrutinized seasonal to inter-annual climate forecasts and recognizes the importance of these products for a variety of applications and for a greater knowledge base even without direct application. USDA expects to make even greater use of 6 to 10 day outlooks, the monthly outlooks, seasonal outlooks, and the inter-annual forecasts as the skill level and accuracy of these forecasts increase.

There is a strong cooperative effort between USDA and NWS. For the past 20 years, USDA has worked side by side with NWS at the JAWF.

Since 1993, the SWRC has been conducting active research on quantifying the effects of ENSO on the precipitation of various regions of the western U.S. SWRC is currently working to evaluate improvement potential for intra seasonal precipitation and temperature forecasts conditioned on a variety of ENSO related indices, *e.g.*, Southern Oscillation Index and Pacific North America Pattern.

<u>Estimates of Agricultural Production</u>

Question 7. Has USDA incorporated estimates of El Niño in its estimates of domestic and foreign agricultural production? If not, will these types of parameters be included in future models?

Answer: No. USDA does not include weather forecasts in its estimates of domestic and foreign agricultural production. These estimates of crop production are based on weather conditions through the date of release. Weather is one factor along with other technological and economic factors. All production estimates assume normal weather from the release date to the end of the growing season. The current situation in Australia is illustrative. Drought is a major factor during the wheat growing season in eastern Australia during an El Niño event. The 1997 El Niño event began after the Australian wheat crop was planted. While some responded to this correlation by reducing the estimated wheat yield in Australia, USDA estimates were stabilized making use of climatological normals until observational evidence during the growing season supported a reaction. Recent rainfall, at a critical growth period, supports USDA's assessment and methodology.

<u>Impacts on US Agricultural Production</u>

Question 8. What is the expected impact of the current El Niño on agricultural production in the U.S.?

Answer: The El Niño is not having a significant impact on the 1997 crop season. The NWS long term outlooks are calling for above normal precipitation across the southern tier of the United states through spring, but the forecasts for June through September 1998 are now calling for near normal temperatures and precipitation.

Some scientists suggest drought occurs in the Midwest during the year following the El Niño onset, but the long-range outlooks do not now indicate this trend.

Question 9. What are the expected regional impacts to U.S. agriculture in California, the South, and the Midwest?

Answer: Excessive rains may occur during the winter months in California, causing some flooding and mud slides. Above-average winter rainfall can be expected across the Gulf Coast region. Warmer than normal winter temperatures often are associated with El Niño conditions in the Midwest. There is also concern in California and Arizona that excessive rainfall could reduce crop prospects for late summer/fall-harvested fruit production (e.g., raisins and citrus). Field crop production has not been affected this year.

Although no one can be certain what the weather will be like this winter, past experience suggests the presence of a strong El Niño increases the probability of an unsettled weather pattern. In the winter of 1982/83, heavy rains hit almost all U.S. winter vegetable areas, affecting planting and harvesting schedules, delaying growth, reducing yields, and affecting produce quality. However, market volume remained near normal with help from larger shipments from Mexico, spurred in part by devaluations of the peso. But with wet weather interrupting the normal flow of product to market, grower and retail prices averaged above the previous year from March through June of 1983.

Impacts on Overseas Agricultural Production

Question 10. Has USDA examined the impact of El Niño on agriculture in developing countries? If so, what are the implications for U.S. food assistance programs?

Answer: El Nino effects have been strongly documented in Central and northern South America, South Africa, and most notably, India and Australia. The effects of El Niño are also being felt or expected to be felt in several developing countries including North Korea, Brazil, Nicaragua, Indonesia, and the Philippines. Agencies involved with food assistance programs are closely following the situation. The effects can not be predicted with certainty, however, as noted in India and Australia. The Indian monsoon, while erratic early, has finished strong and on time. Generous rains in Australia came at the right time for the struggling wheat crop.

The implications for U.S. food assistance programs are expected to be small in the current year. FY >98 requirements may be greater. Funding decisions have already been made, with the only change during the year possibly occurring in the allocation of funds between programs. Funds under P.L. 480's Title II program may be shifted from developmental to emergency programs.

The 1992 El Niño resulted in a severe drought in southern Africa which reduced yields and output by roughly 45 percent. As a result, imports in the region increased to more than 5 million tons—well above the average of 1.5 million tons. Food aid to the region, which usually averages about 1 million tons per year, jumped more than threefold.

Some of the countries in the region are in a better position than they were in 1992 to handle the adverse effects of El Niño. Many countries are aware of impending drought and are already trying to coordinate activities of various organizations within the countries to prepare for it. For example, the government of Zimbabwe has set aside funds targeted specifically for corn imports.

<u>Secondary Impacts of El Niño</u>

Question 11. Much of the focus of El Niño predictions has been on temperature and rainfall. What are some of the secondary effects that have an impact on agriculture, such as pests, and how good an understanding do we have of these effects?

Answer: There are potentially serious effects of El Niño that are not fully understood. Because the El Nino is an erratic event and its impact is difficult to predict, information regarding its impact on pests and disease is anecdotal and after the fact. More research needs to be done.

HEARING OF THE SUBCOMMITTEE ON ENERGY AND ENVIRONMENT
COMMITTEE ON SCIENCE
U.S. HOUSE OF REPRESENTATIVES

on

Preparing for El Niño

Thursday, September 11, 1997

Post-Hearing Questions Submitted to

Mr. Douglas P. Wheeler
Secretary for Resources
State of California

Impact on California

Question 1. How many California counties do you estimate will be affected by El Niño and approximately how many people live in these areas?

Answer: The effect of El Niño on California is not certain. Based on past such events we would expect coastal regions from the San Francisco Bay southward and all of southern California to be strongly affected. The Central Valley region will probably be affected too, especially on the west side, but the impacts are likely to be less. There is high variability in historical impacts; most larger El Niño years have been wet but a few have been relatively dry. We would expect 21 of the 58 counties to be strongly affected. Around 26 million people live in these areas. Another 27 counties in the Central Valley drainage basin are likely to be affected to some degree for a possible total of 48 of the 58 counties. An estimated 6 million additional people live in the Central Valley drainage basin.

Lead Time Need to Prepare for El Niño

Question 2. You note in your testimony that California has broad experience dealing with natural disasters. Timely information is obviously important in mitigating the impacts of severe events. Ideally, how much lead time is necessary for states to prepare adequately for a strong El Niño event and is the lead time provided in current forecasts adequate?

Answer: California's state departments are continually prepared to deal with whatever natural disaster may occur. Given current forecasts for this El Niño event, our preparations have been heightened. Governor Wilson convened an El Niño Summit in early October and has since held eight regional meetings to ensure state/local coordination for winter weather preparedness. (An El Niño Preparedness Regional Briefing Packet is included, for you information.) Obviously, up-to-date forecasts assist the state in focusing on where potential impacts occur; however, currently, state departments are prepared statewide for response to El Niño storm-related impacts.

Accounting for Uncertainty in ENSO Forecasts

Question 3. Clearly, the El Niño forecasts we are capable of providing now, while better than even a few years ago, are still far from perfect. How do state emergency planners treat the uncertainty inherent in any forecast and what is done to minimize the impacts of this uncertainty?

Answer: Emergency planners in general tend to focus on the worst forecast because they want to be as prepared as possible for adverse events. Department of Water Resources Flood Operations staff plan for the maximum event every year, regardless of the forecast. They will run "table top" simulation exercises of potential floods and other disastrous events. They do like to have a well described reasonable range of possible natural events to consider for initial preparation. Actual reaction is keyed more on regular short-range forecasts and warnings and observed events.

Flood Canals in California

Question 4.   One of the challenges of these forecasts pose to federal and state agencies is in the area of flexibility. How does an agency shift resources or change the focus of a program or policy rapidly in response to a forecast of a seasonal event that may come around only every 2 to 10 years. For example, the US Army Corps of Engineers requires permits to clear flood canals. To your knowledge, is anything being done to expedite the permitting processes in the state so that flood canals can be cleared in time for the predicted rains?

Answer: Federal and State agencies cannot be sure of future storm and river flood conditions. They assume, in preparing for the season, that every coming winter can produce a large flood. There has been more urgency this year in view of El Niño based forecasts of heavy rains made as early as last summer and in view of heightened flood awareness from last winter's huge Northern California floods. Actual flood events, of course, are determined by the intensity of specific storms or storm series rather than total seasonal wetness.   A longer wet season does mean more opportunities for flooding, especially in the areas most strongly linked to El Niño wetness in Central and Southern California.

The permitting process for normal and routine channel clearing can be onerous. One goal of the recently convened Governor's El Niño Summit was to obtain clear and consistent channel-clearing permit requirements and procedures from the State and federal governments. The Department of Water Resources has been working with many flood control agencies, the Department of Fish and Game, and the U.S. Army Corps of Engineers to help expedite the State and federal permitting process.

One of the major concerns raised by local flood control channel-maintaining agencies is environmental mitigation. The agencies assert that the cost of mitigation often greatly exceeds the cost of maintenance work. To this end, the State may seek legislative relief at both the State and federal levels to clarify that man-made flood control channels are exempt from permitting in the future.

Energy System in California

Question 5. What are some of the precautions the California Energy Commission is making to make the energy system disaster-proof?

Answer: The system is already designed and operated to maintain high level of reliability even if disasters of one type or another affect individual components of the system. Additionally, the California Energy Commission (CEC) has developed an Energy Shortage Contingency Plan, which coordinates energy emergency planning with state and local agencies, other states, the federal government and private industry. As part of the plan, the CEC maintains a network of public and private sector contacts to ensure that a communication system is in place in response to severe weather. In the event of an energy supply shortage or disruption due to severe weather, the CEC has developed conservation strategies which can be applied on either a voluntary or mandatory basis.

Interagency Research into Hazard Mitigation

Question 6. Do you believe, as has been suggested, that a federal interagency program of hazard mitigation research is a worthwhile idea?

Answer: Yes. This would expand the range of hazard mitigation concepts and measures for reducing risk and losses from natural hazards and improve coordination among federal programs. Hopefully, this would also lead to federal policy changes to broaden the scope of hazard mitigation alternatives eligible for federal grant funding (not just based on federal benefit or federal cost containment). Also, if implemented, it would be extremely important that State and local communities be given an opportunity to provide suggestions and/or be in the review/comment process.

Response of Local Governments

Question 7. How have local and city governments in California responded to El Niño, and what are they doing to mitigate its impacts?

Answer: On an annual basis, regardless of an El Niño forecast, local and city governments undergo winter weather preparedness, especially as a follow up to last years storms that caused extensive damage to many areas. These activities include updating of emergency plans, storm channel clearing, and berm and levee re-inforcement.

Additionally, The Governor's Office of Emergency Services maintains local offices which have conducted Alert and Warning Workshops as well as a series of workshops on recovery and response to severe winter weather. The Office of Emergency Services and the Department of Water Resources have also conducted eight regional El Niño Preparedness workshops to support local jurisdictions and enhance state/local coordination.

ENSO and Agricultural Production in California

Question 8. What are some of the expected impacts of El Niño on California's agricultural production?

Answer: For 50 consecutive years, California has been the number one agricultural state in the nation. Thus, impacts from El Niño could greatly impact California's agricultural production given that nearly one-third (29 million) of California's 100 million acres of land are devoted to this industry.

In an El Niño year, weather conditions tend to vary widely across the state and it is unclear as to where specifically the impacts will occur. The general prediction is that southern California will be receive above average rainfall and that northern California may or may not see above normal amounts of precipitation. If above normal rainfall occurs, dry farmed agriculture, including unirrigated grazing land and pasture, should benefit from El Niño. It is also likely to be a good year for water supply and ground water recharge. Spring conditions are more important for most irrigated crops. However, potential adverse impacts to agriculture could occur from excessive rains in February and March. These impacts may include actual storm damage to trees or crops, and/or flooding which can lead to disease. A wet spring could also delay ground preparation for annual crops.

Positive Impacts of El Niño in California

Question 9. There are sectors in California, such as sport fishing, that El Niño is expected to benefit. How is California poised to take advantage of the positive impacts of El Niño?

Answer: Because of the warmer ocean temperatures, several types of fish, such as Yellowfin Tuna and Albacore are being caught as far north as off the coast of Eureka in northern California. The results of this include an increase in the number of boats and recreational fishing days which corresponds to in increase in receipts for commercial fishing vessels, an increase in the number of fishing licenses sold and other related effects. The actual positive economic benefits for the sport fishing industry cannot be quantified until after the end of the fishing season.

<u>Applying the Lessons of the 1982-1983 El Niño</u>

Question 10. What are some of the lessons California resource managers learned during the 1982-83 El Niño event and how have these lessons been applied to prepare for the current event?

Answer: Climate forecasters have been warning that 1982-83 is a likely pattern for this coming water year. That was the wettest water year this century, including a huge snowpack which also caused snowmelt flooding in the San Joaquin Valley. Foothill multipurpose reservoir operators did a very good job that year of limiting downstream river flows within the regulation capacity of the system. The spring and early summer runoff forecasts produced by our Cooperative Snow Surveys program were and are a vital part of the reservoir and flood control system regulation process. For much of northern California, 1982-83 was a long wet rainy season with 8 consecutive wet months from October through April. But the year did not produce large rain floods on the major rivers (upper Sacramento excepted).

While the 1982-83 El Niño caused extensive damage to coastal areas, those areas have taken actions to improve preparedness to minimize damage from waves. In inland areas, we learned where we might expect flooding due to excessive snow melt runoff and much has been done to strengthen levees the Sacramento-San Joaquin flood control system, including the Delta.

El Niño PREPAREDNESS

California Resources Agency

California Department of Water Resources

O E S
Governor's Office of Emergency Services

El Nino Preparedness

Regional Briefing

El Niño PREPAREDNESS

California Resources
Agency

California Department of
Water Resources

Governor's Office of
Emergency Services

El Nino Regional Briefings

1.	**Long Beach -- Friday, Oct. 24** -- 9.30 a.m. until noon, Long Beach Convention and Entertainment Center, Center Theater, 300 East Ocean Blvd.

2.	**San Diego -- Monday, Nov. 3** -- 9 a.m. until noon, State of California Building, Conference Room B109, 1350 Front Street.

3.	**Chico -- Monday, Nov. 10** -- 9:30 a.m. until noon, Chico Municipal Center, Council Chambers, First floor, 421 Main Street.

4.	**Riverside -- Thursday, Nov. 13** -- 9:30 a.m. until noon, Riverside Convention Center, Rain Cross Ballroom, 3443 Orange Street.

5.	**San Jose -- Friday, Nov. 14** -- 9:30 a m. until noon, George Shirakawa Community Center, Black Forrest Room, 2072 Lucretia Avenue.

6.	**Santa Barbara -- Wednesday, Nov. 19** -- 9:30 a.m. until noon, City Hall, Council Chambers, 2nd floor, 735 Anacapa Street. (La Guerra Plaza).

7.	**Santa Rosa -- Monday, Dec. 1** -- 9:30 a.m. until noon, Veterans Memorial Building, Dining Room, 1351 Maple Avenue.

8.	**Fresno -- Thursday, Dec. 11** -- 9:30 a.m. until noon, City Hall, Council Chambers, 2nd floor, 2600 Fresno Street.

Appearances of El Niño have been documented as far back as 1726. In the winter, the Equatorial Countercurrent shifts southward from the Gulf of Panama to the coast of South America, bringing warm water into the normally cold waters and forming a warm current. Derived from Spanish words meaning "the child," the term El Niño was originally coined in reference to these warm currents arriving annually during the Christmas season off the coast of Peru and Ecuador. The term was later restricted to the particularly strong warming that disrupted the fish and bird populations, occuring when the westward trade winds blowing across the tropical Pacific Ocean weaken and allow a mass of warm water to slosh eastward toward South America instead of Australia. This warm water mass in turn heats the atmosphere and other parts of the ocean, resulting in water temperatures off California's coast up to 5 degrees above normal and alteration of storm tracks.

Scientists generally refer to El Niño and its related phenomena as the El Niño-Southern Oscillation (ENSO). A major ENSO disturbance affects other ocean-current patterns and causes climatic changes worldwide, resulting in disastrous flooding and drought in some areas. In 1982-83, an ENSO caused droughts in India, Indonesia, Australia and the Philippines and flooding in Peru and Ecuador. In California, 1982-83 was marked by more than 26 inches of rain and over $100 million in coastal losses, including destruction of 33 ocean front homes, damage of another 3000 homes and 900 businesses and $35 million in damages to coastal public recreation facilities.

Oceanographic and meteorological indicators show that strong El Niño conditions have been developing in the tropical Pacific Ocean since Spring 1997, and the World Climate Research Programme of the United Nations predicts that the 1997-98 El Niño could be "the climatic event of the century". Experts are comparing this year's El Niño to that of 1982-83, and some climatologists warn that El Niño is already so extreme and affecting so large an area in the Pacific Ocean off South America that it could, as Nicholas Graham of Scripps Institution of Oceanography in La Jolla, California, says, "cause billions and billions in damage worldwide".

Although the weather phenomenon may have created better fishing conditions for some Northeast fisherman, El Niño may bring increased storminess to California, marked by flooding and coastal damage. The likelihood of a wet winter season is high in the South Coast region, with coastal and lower elevation areas being more likely to have flooding, especially south of San Francisco. Ocean levels will be higher than normal with increased risk of wave damage during high tide and increased levee problems in the Delta. Additionally, rains are likely to start earlier in the fall; once soils are saturated there will be greater risk of landslide movement and flooding, with historical problem areas of land movement becoming reactivated. As this year's warming is earlier than in 1982, the current conditions are without precedent.

Attachment 1

El Niño Preparedness Needs by Region

Region	El Niño Forcast	Special Conditions
North Coast	*near normal*	Monitor high tides and waves, prepare river flood forecasts as needed.
San Francisco Bay	*normal to wet*	Monitor high tides and waves. Be aware of landslide threats during heavy storms once soil is saturated.
Central Coast	*wet*	Monitor high tides and waves, some river flooding possible.
South Coast*	*wet*	Monitor high tides and waves, some stream flooding possible. Be aware of landslide threats during heavy storms once soil is saturated.
Sacramento River*	*normal to wet*	Monitor status of levee repairs, prepare river flood forecasts. Develop predeployment strategy for forecasted high water events.
Delta*	*normal to wet*	Closely monitor impacts to levees from high water and high tides.
San Joaquin River*	*normal to wet — potential large snowmelt flood in spring*	Monitor status of levee repairs, prepare river flood forecasts. Develop predeployment strategy for forecasted high water events.
Tulare Lake*	*normal to wet — potential large snowmelt flood in spring*	Prepare river runoff forecasts. Develop predeployment strategy for forecasted high water and Tulare Lake flooding.
North & South Lahontan (Eastern Sierra)	*normal to wet*	Monitor conditions, including snowpack and snowpack runoff.
Colorado Desert	*wet*	Monitor Colorado River runoff forecasts and storage.

*Regions requiring a higher level of preparedness

Observed Sea Surface Temperature Anomaly (Deg C)

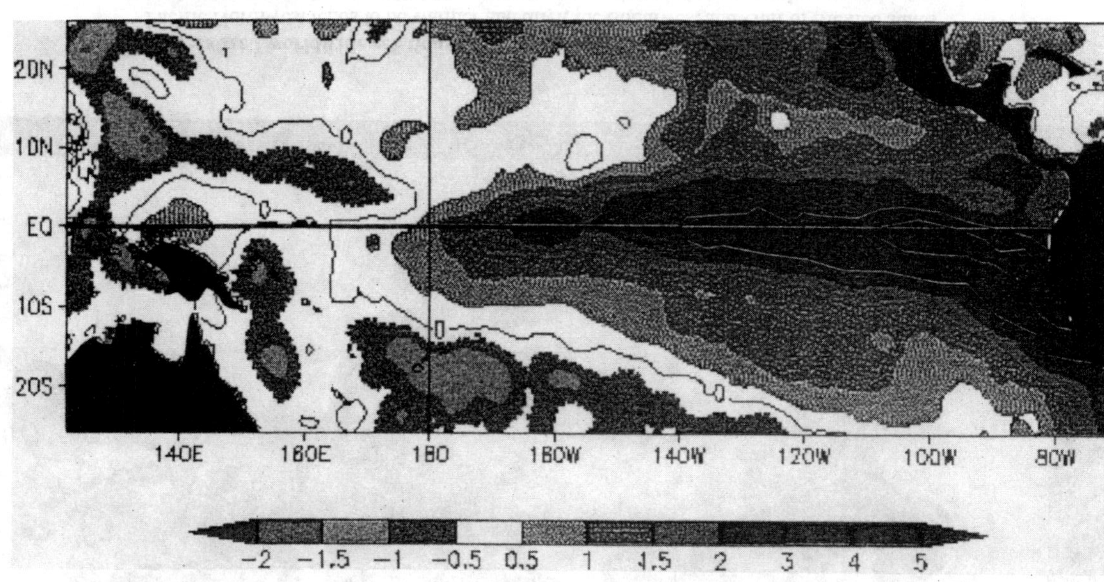

7-Day Average Centered on 27 August 1997

Sea surface temperature anomaly map of the tropical Pacific Ocean for the last week in August, in degrees Celsius. This shows how the temperature of the sea surface compares to average (the anomaly). A large warm area extends from South America to the international date line, with a region 4 degrees above normal on the equator and off the coast of Peru. Note that the western Pacific is slightly cooler than normal.

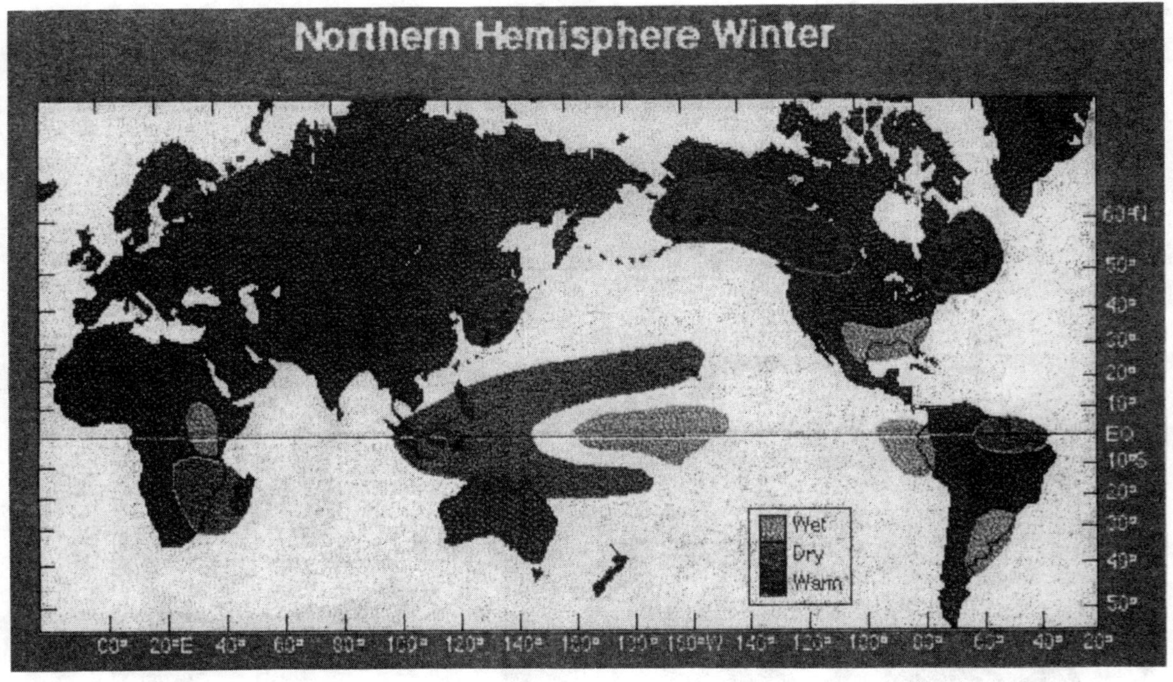

Generalized worldwide relationships during our winter for El Niño years. The Pacific Northwest tends to be warmer and drier; the Southwest and Gulf of Mexico states tend to be wetter. In strong El Niño years the wet region tends to expand westward into Arizona and Southern California and northern Mexico.

STATUS REPORT OF FEAT RECOMMENDATIONS
AUGUST 26, 1997

	FEAT EMERGENCY RESPONSE RECOMMENDATIONS	STATUS
1	**Improve Local Maintaining Agency Emergency Response Coordination and Operations:** Direct OES to develop and test guidelines that clarify how federal, State, and local agencies will coordinate joint field emergency operations under Standardized Emergency Management System. The Guidelines should integrate local agencies that maintain levees and flood control structures into the overall emergency response organization. These guidelines must define fiscal responsibilities, emergency response, and statutory and regulatory authorities.	OES is currently conducting a series of meetings with local governments, DWR, and local levee maintaining agencies to develop agreements and concepts that will become the basis for future guidance. OES is developing a three-part guidance document. Part 1 will describe how state and federal agencies with field operations can coordinate their actions with local government under SEMS and use organizational tools, such as Area Command, that have already been developed for use in the incident Command System. Part 2 will demonstrate how the organizational tools defined in Part 1 may be specifically applied in a flood scenario. Part 3 will delineate how resource allocation and fiscal responsibility will be handled by the various participating agencies.
2	**Local Maintaining Agency Emergency Plans:** Encourage local agencies responsible for maintenance of levees, and flood control structures, to coordinate an emergency plan and response actions with the appropriate city and county emergency management agency.	OES is currently conducting a series of meetings with local governments, DWR and local levee maintaining agencies to develop agreements and concepts that will become the basis for these plans and procedures. OES will develop a model emergency response plans and procedures for levee maintaining agencies. These plans and procedures will specifically address how levee maintaining agencies, as a part of local government under SEMS, can be integrated into the local emergency management organization of the city or county where they operate.
3	**Model Emergency Procedures:** Direct DWR, in coordination with OES, to develop model emergency procedures and training for use by local maintaining agencies in development of local plans.	OES is developing model emergency procedures and training for use by local agencies
4	**Alerting and Warning Exercises:** Direct OES and DWR to conduct flood emergency workshops annually, prior to the flood season. This effort will focus on the dissemination of critical information to decision makers and the effective use of tools for conveying emergency information to the public in a timely manner.	OES has initiated intensive training for local agencies directed at improving information dissemination during floods. OES is currently clarifying authorities for evacuation and developing a guidelines document. Workshops will be conducted throughout the state . Authorities will be identified and a guidelines document for evaluation will be developed.
5	**Improve Evacuation Procedures for Mobile Home Parks in Floodways:** Direct OES in Corrd w/HCD to review mobile home and RV park evacuations during the 1997 floods and to improve evacuation future floods.	A draft analysis and model plan have been developed. An analysis is being developed that will identify the responsibilities for evacuation planning and plan review at mobile home and recreational vehicle parks. In addition, a model evacuation plan for these parks will be developed.

6	**Livestock and Pet Evacuation:** Direct OES in cooperation with local animal control officers, the Department of Food and Agriculture, and UC Cooperative Extension, to review procedures for livestock and pet evacuation and develop animal safety and relocation procedures to be used in future emergencies.	Meetings are being held between the designated veterinary coordinators and the local emergency management communities. Existing emergency plans that include an animal or livestock element have been surveyed. The intent is to integrate the animal evacuation system developed by the California Veterinary Medical Association into the broader emergency management system. A model local government planning guidance for addressing animal and livestock issues will be developed.
7	**Sacramento San Joaquin Delta Waterways:** Direct OES and the Department of Boating and Waterways, in cooperation with the U.S. Coast Guard and the Delta Protection Commission, to develop a plan of action for future emergency closures of the Sacramento San Joaquin Delta waterways to non essential vessel traffic during periods of extremely high waters.	Coordination with the U.S. Coast Guard an the Department of Boating and Waterways to clarify legal authorities is taking place. Protocols will describe the process and responsibilities for closure of the Delta Waterways.
8	**Response Information Management System (RIMS):** Directs OES to explore the feasibility of developing RIMS for application to local governments which currently do not have access to it.	RIMS software and training are being made available to counties and state agencies. This will result in an expanded application of RIMS.
9	**DWR Emergency Management Function:** Direct DWR to establish a emergency management function to better meet the requirements of the State's Emergency Services Act and the Standardized Emergency Management System. More emphasis should be placed on advance planning for all types of emergencies, and year round coordination with OES and other local, State, and federal responding agencies.	DWR is still in the process of seeking the necessary approvals for its emergency response manager. DPA has approved the position and it is now pending with the State Personnel Board. The position has been advertised in anticipation of its approval.
10	**Disaster Assistance Funding Guidance:** Direct OES to use the Standardized Emergency Management System maintenance system to provide guidance on disaster assistance funding. This includes developing guidelines and training that clarify the responsibilities and benefits of emergency proclamations and declarations. To support this effort, OES will also develop a federal and State disaster assistance program matrix describing types of assistance provided, application requirements, time frames, and restrictions.	A preliminary version of the computer based system is currently under peer review. OES wills produce a matrix outlining the process for requesting the types of disaster assistance available under various emergency proclamations and declarations. This matrix will be computer-based and available on the web to allow interactive linking to web sites providing more detailed disaster assistance information. The matrix will also identify responsible state or federal agency(s) and governing authorities. OES will utilize ongoing regional and local government coordination meetings to provide training on this guide to disaster assistance funding.

11	**Flood Center Event Tracking and Computer:** Direct the Department to assure that flood event tracking and reporting systems are completed, maintained, and staffed, including training of staff used only in emergencies.	Most of this activity, $360,000 was covered in recently chaptered SB 4x (Costa). A small amount for one time computer purchases $90,000 as eliminated during the final budget conference committee negotiations.
12	**Multi Party Agreement on Payment:** Directs OES to coordinate, consistent with FEMA guidelines for reimbursable costs, a multi party agreement among affected parties addressing payment for flood emergencies and pre emergency flood response.	As a result of the January 1997 storms, OES developed an accelerated payment process. This process included the expedited handling and approval of emergency response damage survey reports, combined processing of federal and state claims, automatic payment of federal and state cost shares for emergency response costs, and a reduction in the number of state forms. OES has implemented an automated ledger system which has substantially reduced the processing time for federal disaster assistance. OES will develop written guidance that identifies emergency response and recovery costs which are generally reimbursable and facilitate expedited payments to local agencies for the cost of emergency response operations. The guidance will include a model for levee maintaining agencies to seek payment of flood emergency costs consistent with FEMA requirements.
13	**Authority to Fund Capital Outlay:** Recommends that the Legislature give the Department of Finance the authority to allow the use of Section 8690.6 to allocate funds for capital outlay projects needed to maintain essential State functions and/or to ensure public safety in the event of future disasters.	This authority is in AB 10x (Machado), and should be heard in the Senate Governmental Organizational Committee this week.
14	**Expand and Adequately Fund Long Term Stream Gage Database:** Urge the U.S. GExecutive Orderlogical Survey to expand its surface water data collection program and support long term record of flows for gaging stations for more rivers and streams in California. This database is needed to define the hydrology of watersheds and provide statistics for critical water use determinations and to more accurately determine floods of a specific frequency, particularly the 100 year event which is the basis of NFIP floodplain mapping.	The USGS has responded to Secretary Wheeler's letter. Unfortunately the response indicated that he Federal Government cannot be counted upon due to limitations n federal funding priorities. DWR will continue to pursue a better response
15	**Uniform Flood Frequency Determination and Single Elevation Datum:** Urge federal agencies to standardize flood frequency determinations and adopt a single elevation datum using English units.	The USGS has responded to Secretary Wheeler's letter. The response is non-committal. DWR will continue to pursue a better response.

AGENCY & DEPARTMENT
EL NIÑO PREPAREDNESS ACTIONS

- Business, Transportation & Housing Agency

- Governor's Office of Emergency Services

- Employment Development Department

- California Environmental Protection Agency

- Department of Finance

- California Department Food & Agriculture

- California National Guard

- California Public Utilities Commission

- Resources Agency

- Youth & Adult Correctional Agency

BUSINESS, TRANSPORTATION
& HOUSING AGENCY

California Department of Transportation (CalTrans)

Actions Completed for Emergency Preparedness:

CalTrans cleaned more than 30,000 storm drains statewide and completed more than $200 million repairing storm damage from the winter of 1996/1997.

CalTrans purchased more than 295,000 tons of abrasives and 12,000 tons of deicers at a cost of $3.5 million to melt snow and improve traction on mountain roads.

CalTrans repaired flood damage to Route 50 and facilitated the rapid re-opening of Route 50 in early 1997.

CalTrans, in conjunction with the Department of Conservation, identified potential slide locations on Route 50. Forty locations were identified for stabilization of slide material, armoring of the river bank, replacement and upgrading of drainage facilities and roadway paving. The contracted work should be completed by mid to late October 1997.

CalTrans installed slide monitoring equipment at 5 potential slide locations along Route 50. Response protocol for landslide activity was established through a 24-hour dispatch location.

CalTrans established the Route 50 Coalition of county officials, homeowners and businesses to meet periodically with the department to discuss plans, strategies and impacts of disasters on the community.

Actions Taken to Prepare for El Niño:

CalTrans will coordinate with county Office of Emergency Service (OES) offices and other emergency service providers to provide emergency response to floods and slides, and the department has updated its emergency response plans internally and with local response agencies.

Cal Trans, in conjunction with the California Highway Patrol and OES, has developed communications and decision-making protocols in the event that parts of the state highway system are closed or used to manage a storm-related emergency.

CalTrans maintains a fleet of 8,000 pieces of equipment in a state of readiness to respond immediately in the event of any weather-related storm damage.

Public Information Contact:

Jim Drago
916/654-4677

GOVERNOR'S OFFICE OF
EMERGENCY SERVICES

Actions Completed for Emergency Preparedness:

The Governor's Office of Emergency Services (OES) conducted Alerting and Warning workshops to ensure that emergency information is conveyed to the public in a timely and effective manner.

OES developed Evacuation Guidelines that identify authorities and responsibilities for flood evacuations that will assist local governments in emergency response.

OES conducted flood awareness and preparedness workshops throughout the Inland Region with levee maintaining agencies and local governments. These workshops identify the methods and procedures to obtain assistance and maintain a clear communication during a potential flood fight.

OES developed guidance to enhance SEMS Guidelines regarding how state and federal agencies with field operations can more effectively coordinate their actions with local governments under SEMS and use of organizational tools, such as Area Command, that have already been developed for use in the Incident Command System (ICS). Guidance also addressed resource allocation improvements and the fiscal responsibilities of the various participating agencies.

OES developed draft emergency planning guidance for levee maintaining agencies which is in final review. Distribution and training on the guidance is scheduled for October 1997.

OES developed emergency planning guidance for mobile home and RV parks in conjunction with the Department of Water Resources (DWR). This guidance includes a model plan for mobile home parks, with emergency information for residents.

OES conducted, with DWR, joint planning sessions to develop methods for SEMS to conduct multi-jurisdictional flood fighting operations. The procedures are in final review and will be ready for use by the end of October 1997.

OES expanded the Response Information Management System (RIMS). OES has connected 15 State agencies, 52 counties, FEMA Region IX, the American Red Cross HQ in Long Beach and the United States Coast Guard HQ in Alameda. OES projects to have the remaining 6 counties connected by the end of October 1997.

OES developed a levee status database that will facilitate the flow of critical levee information from DWR to county EOCs.

OES provided introductory SEMS training at a meeting of the California Animal Control Directors' Association. OES has also developed a model Memorandum of Understanding to be used by local governments in their emergency planning that will include animal care agencies.

OES developed, in cooperation with the state Department of Boating and Waterways, the U.S. Coastguard and Delta Commission, a protocol for emergency boating restrictions and closure of waterways. This will provide a timely response to the need to restrict nonessential boating activities during flood fight efforts.

OES developed, and is testing, an interactive computer-based matrix outlining the process for requesting financial assistance and the types of assistance available under various emergency proclamations and declarations. The matrix will identify responsible state or federal agencies and governing authorities. It will be available in October 1997 and available in an electronic format on RIMS by the end of the year/early 1998.

OES initiated a process to facilitate expedited payments to local agencies for the cost of emergency response operations including a model for levee maintaining agencies to seek payment of flood emergency requirements consistent with FEMA requirements.

Actions to be Taken to Prepare for El Niño:

All three OES regions will address flood awareness activities for the next Mutual Aid Region Advisory Committees (MARACs). The materials to be discussed were developed in response to the FEAT Report, and the Governor's El Niño Executive Order will be shared and discussed with regional representatives of state agencies and local governments.

OES California Specialized Training Institute (CSTI), in late October 1997, will be conducting a specially designed flood scenario emergency maintenance management class oriented toward local governments to include levee maintaining agencies.

OES is continuing to improve its ability to deploy Geographic Information System (GIS) teams to support incident command posts.

OES is developing written guidance that identifies emergency response and recovery costs which are generally reimbursable.

OES is conducting workshops with local governments to provide training and guidance on methods for recovering costs associated with flood emergencies.

OES is conducting its Annual Winter Storm Awareness Campaign from October 27-31. The office is working with other state agencies to link critical disaster information available to the general public on the Internet.

Public Information Contact:

Tom Mullins
916/262-1843
916/262-1832 (Emergency News Center)

EMPLOYMENT DEVELOPMENT DEPARTMENT

Actions Completed for Emergency Preparedness:

During the flood of January 1997, the Employment Development Department (EDD) Job Service field offices collected information on clean-up services, food banks, shelter and other local services available to displaced victims, which was distributed locally by EDD staff.

Actions Taken to Prepare for El Niño:

The EDD Information Security Office (ISO) is the primary liaison with the Office of Emergency Services (OES), securing and disseminating information for EDD related to the emergency. ISO will be available to coordinate as a sister agency to the Department of Social Services on the staffing of Regional Emergency Operational Centers which provide care and shelter during an emergency during to people who have been displaced from their homes, and ISO helps mobilize cleanup crews to specific sites through the Job Service casual labor programs.

EDD administers the federal Disaster Unemployment Assistance (DUA) program in California. The automated DUA program will provide financial assistance and employment services to the self-employed and other jobless workers not eligible for regular unemployment insurance (UI) when they are unemployed as a direct result of a major natural disaster. There is no waiting period for DUA as there is for regular UI claims. EDD forwards a Governor Action Request to waive the waiting period for regular UI as a direct result of a disaster.

If the Governor declares a state of emergency, the Department may extend the time requirements for filing employer tax returns or reports and the payment of employer and worker. The extension may be granted to employers affected by a disaster in a state of emergency. This extension has been generally granted to requesting employers.

EDD will work closely with FEMA and the Office of Emergency Services to respond to disasters with Job Training Partnership Act (JTPA) funding from the Department of Labor to administer disaster recovery projects which feature temporary job creation. Work conducted may include assisting in clean-up, repair and reconstruction of public and private non-profit property; prevention, education and public safety awareness efforts; and supplementing existing staff whose workloads and caseloads have increased as the result of the disaster.

Public Information Contact:

Suzanne Schroeder
916/654-9029

CALIFORNIA ENVIRONMENTAL PROTECTION AGENCY

Actions Completed for Emergency Preparedness:

Cal-EPA conducts quarterly Emergency Response Multi-Agency Coordination (ERMAC) meetings, to share information, coordinate, ensure consistency and set priorities for emergency management through the agency's organization. El Niño will be the focus of the October 1997 meeting.

In January 1997, the Integrated Waste Management Board (IWMB) adopted an integrated disaster debris recovery plan, which serves as a guide to local jurisdictions preparing its own, locally specific plans. The plan was presented to local representatives in five statewide workshops.

In July 1997, the IWMB issued an advisory, providing guidance to local agencies on the issuance of emergency waiver of solid waste facility standards to facilitate disaster cleanup efforts.

The Department of Toxic Substances Control (DTSC) established zone contracts to provide hazardous materials emergency response support, which included communications mechanisms to coordinate the handling of hazardous materials response work with federal agencies.

DTSC centralized management of flood response activities in its Department Operating Centers. The department has also actively pursued FEMA reimbursement to be available for potential 1998 flood work.

The State Water Resources Control (SWRCB) Board requested all public-owned treatment works to update and submit their wet weather contingency plans.

The SWRCB has increased inspections and dairy industry awareness of runoff problems. The department has facilitated municipal stormwater channels being cleared of vegetation and Regional Board staff are expediting related permit processing. They have also inspected landfills prior to the wet weather season to ensure erosion control.

The Department of Pesticide Regulation will work with local agricultural commissioners to ensure farming operators are advised of the best management practices regarding the storage and handling of pesticide inventories.

The Office of Environmental Health Hazard Assessment developed a 24-hour response system for providing emergency toxicology and risk assessment assistance in the event of a hazardous chemical spill or release.

Actions Taken to Prepare for El Niño:

Cal-EPA has prepared Permit Assistance Centers to allow for an integrated rapid response to support disaster recovery efforts.

The IWMB has prepared to assist OES on disaster debris disposal, recycling and recovery of materials and special handling and collection needs for hazardous household materials.

The Air Resources Board is ready to work with local air districts to expedite burning of flood debris, as required.

Public Information Contact:

Joe Irvin
916/324-9670

DEPARTMENT OF FINANCE

Actions Completed for Emergency Preparedness:

The Governor signed legislation to appropriate $5.4 million for planning and advanced deployment of resources. These funds would be allocated by the Department of Water Resources to other agencies as needed to plan for possible flooding and will allow for deployment of staff to areas of particular concern in the event that flooding is forecast. In the past, such deployments have often occurred only after floods were in progress.

Disaster Funding Mechanisms:

If the El Niño were to result in serious flooding, there are a number of mechanisms which would be used to fund necessary response efforts by state and local governments, to assist victims and to repair damaged public facilities. Severe flooding, such as that occurring in 1995 and 1997, would normally result in disaster declarations by the Governor and at the federal level. State agencies respond to the disaster as directed by the Governor, and a federal declaration makes the state eligible for federal assistance of various sorts.

Under current law, the Governor is authorized to allocate funds to state agencies by Executive Order in the event of a declared disaster. This mechanism is typically used to fund response costs by state agencies as well as funding provided to local agencies through the Office of Emergency Services.

Victims of flooding would also be eligible for tax relief. Current income tax law and bank and corporation tax law allow non-business casualty losses over $100, not reimbursed by insurance, to be deducted if the loss for the year exceeds 10% of adjusted gross income. Casualty losses on business property are not subject to the $100 and 10% of adjusted gross income limitations that apply to non-business property. Fifty percent of unused losses may be carried forward for up to 15 years as a net operating loss. Casualty losses that occur in a federally declared disaster area may be claimed in the year that the disaster occurred, or the preceding year, which allows disaster victims to immediately take advantage of these provisions. In the aftermath of disasters, legislation has also often been enacted to authorize victims to carry forward 100% of any unclaimed losses for up to five years, with 50% of any remaining losses carried forward for an additional 10 years.

Current law also provides that a county board of supervisors may adopt an ordinance authorizing an assessee to apply for the re-assessment of property damaged in a disaster, and that the property owner may apply to the county for deferral of the property tax until the next installment due following the disaster. The county may apply to the state for a "bridge loan" to cover cash flow losses during the period of deferment.

Public Information Contact:

H.D. Palmer
916/323-0648

CALIFORNIA DEPARTMENT
OF FOOD & AGRICULTURE

Actions Completed for Emergency Preparedness:

California Department of Food and Agriculture (CDFA), with the Office of Emergency Services (OES), has established procedures to coordinate the evacuation and relocation of livestock.

CDFA, will assist the Department of Pesticide Regulation and County Agricultural Commissioners (CACs) with the communication of best management practices regarding the storage and handling of pesticide inventories.

Actions Taken to Prepare for El Nino:

CDFA will coordinate disaster information with CACs, County Farm Bureaus, and commodity and industry groups.

CDFA will work in conjunction with OES, the California Veterinary Medical Association, CACs, the California Animal Control Directors Association, the University of California (UC) at Davis Veterinary School, and the UC Cooperative Extension Service concerning the evacuation and relocation of livestock and pets.

CDFA will facilitate the communication of disaster/emergency information to the agricultural community through CDFA's website/home page at http://www.cdfa.ca.gov.

CDFA will participate in the OES Coastal Region Mutual Aid Regional Advisory Committee, which will facilitate the formalization of roles and responsibilities, both public and private, to ensure animal safety during disasters.

Public Information Contacts:

Kevin Herglotz
Katheryn Carlisle
916/654-0462

CALIFORNIA NATIONAL GUARD

Actions Completed for Emergency Preparedness:

The California National Guard continues to monitor critical equipment on a monthly basis for emergency support.

The National Guard has practiced search and rescue and has trained other agencies in air loading procedures.

Actions Taken to Prepare for El Niño:

The National Guard has planned for early deployment of planning and command and control staffs, since early staging will decrease response time and improve support to civilian authorities. In addition, the National Guard has planned on simultaneous commitment to more than one regional emergency situation.

The National Guard is exploring the use of military technology to enhance relief and search and rescue efforts.

Public Information Contacts:

Lt. Colonel Doug Hart
916/854-3829

Lt. Colonel Kevin Ellsworth (Legislative Issues)
916/854-3534

CALIFORNIA
PUBLIC UTILITIES COMMISSION

Actions Completed for Emergency Preparedness:

The Public Utilities Commission (PUC) has required and reviewed emergency plans for electric and water utilities, covering internal and external communications, use of utility-wide resources, mutual aid agreements with other utilities and consumer and press information centers.

The PUC has ordered upgrades for overloaded customer call centers which were overloaded in previous storms.

The PUC has tightened tree-trimming requirements, to minimize outages due to contact between trees and electric transmission lines.

The PUC has established standards for inspection, record-keeping, and repair of electric utility distribution systems, to assure that they are well-maintained and ready for severe weather.

The PUC has met with utilities and the new Independent System Operator, the CEC, and OES to establish new emergency communications links. The Commission has also established emergency communication links with the electric and water utilities.

The Commission has established a expedited procedure to consider disaster-related costs incurred by utilities.

Actions Taken to Prepare for El Niño:

We are asking utilities to assure that their emergency resources are a full strength during the storm season.

We are asking the utilities to test communications links with the Independent System Operators as well as OES emergency operations centers at the state, region, and county levels.

We are simplifying communications links between the electric utilities, the Commission, the CEC, and other state agencies.

Public Information Contact:

Dianne Dienstein
415/703-2423

RESOURCES AGENCY

Actions Completed for Emergency Preparedness:

The Resources Agency has developed an El Niño website on the California Environmental Resources Evaluation System (CERES) at http://ceres.ca.gov/elnino. The website is geared for current information, forecasts and predictions, background information, maps, spatial data, past events and educational resources.

Departments of the Resources Agency have completed or are taking the following actions:

Department of Water Resources (DWR)

Actions Completed for Emergency Preparedness:

DWR has worked with local Reclamation Districts, the US Army Corps of Engineers and the state Reclamation Board on levee repair issues.

DWR has added almost 50 sites identified in the January 1997 floods to the telemetry network to improve flood forecasting operations.

DWR conducts flood fighting classes to help equip volunteers, California Conservation Corps, local officials and Reclamation District staff with correct techniques for fighting floods and making emergency repairs to levees. DWR will have trained 900 people as floodfighters by the end of the year. Information includes how to protect homes, protect beach front areas, protect slopes, fill sandbags, elevate homes in flood-prone areas and protect levees.

Actions to be Taken to Prepare for El Niño:

DWR provides river forecasts, in cooperation with the National Weather Service.

DWR is assisting OES in preparing media outreach during the Annual Winter Storm and Flood Preparedness Week.

DWR will train public information officers in October to staff the Flood Operations Center, which is activated in Sacramento during severe storms and high water threats and can function as an interagency 24-hour Emergency Operations Center.

Public Information Contact:

Anita Garcia-Fante
916/653-7431

Department of Boating & Waterways (DBW)

Actions Completed for Emergency Preparedness:

On August 19, 1997, the DBW co-sponsored an El Niño Workshop for local agencies at Scripps Institution. The department maintains an oceanographer at Scripps to monitor weather and climate phenomenon.

DBW conducted interviews with marina and reclamation district staffs to examine events associated with the January 1997 levee breaks which caused portions of marinas to be swept away and into two Delta bridges. Site visits were also made and structural calculations and drawings of affected marinas were reviewed.

DBW administers an emergency loan program for storm damage of boat marinas operated by public agencies.

On October 2, 1997, DBW met with representatives of the department to plan a workshop, including such agencies as the US Coast Guard, Office of Emergency Services and Department of Water Resources to develop an action plan for limiting use of the Sacramento-San Joaquin Delta during flood conditions.

Actions to be Taken to Prepare for El Niño:

After DBW has finished collecting technical data from piling suppliers, design consultants, and other State agencies on the design of steel and concrete piles for use in the Delta, the department is planing to promulgate a new Pilings Manual standard.

DBW will continue to coordinate the action plan for the Delta with other federal, state and local agencies.

Public Information Contact:

Megan Standard
916/322-1819

California Energy Commission (CEC)

Actions Completed for Emergency Preparedness:

The CEC has developed the Energy Shortage Contingency Plan, which coordinates energy emergency planning with state and local agencies, other states, the federal government and private industry. As part of the plan, the CEC maintains a network of public and private sector contacts to ensure that a communication system is in place in response to severe weather.

In the event of an energy supply shortage or disruption due to severe weather, the CEC has developed conservation strategies which can be applied on either a voluntary or mandatory basis.

The CEC has developed an Emergency Fuel Allocation Program. During a disaster, the commission will provide support to the OES by coordinating the fuel supply for emergency vehicles. The commission has also developed a program to provide emergency fuel through a priority distribution system, if market forces and voluntary conservation are unable to provide adequate fuel due to a prolonged and widespread shortage.

Public Information Contact:

Claudia Chandler
916/654-4989

California Conservation Corps (CCC)

Actions to be Taken to Prepare for El Niño:

The CCC has a strong presence in Southern California as well as the rest of the state and will provide at least 50 crews (more than 800 corpsmembers and staff) trained in floodfighting and flood recovery work to assist local, state and federal agencies across the state. Word can include levee reinforcement, sandbagging to protect home, businesses and public structures, debris removal, evacuation assistance, levee patrol and hillside stabilization. Crews are available 24 hours a day, seven days a week.

The CCC can continue to provide trained staff to assist at Regional Emergency Operations Centers throughout the state.

The CCC continues to provide corpsmembers to assist with flood channel clearance and flood prevention work. Work is currently taking place in Humbolt, Kern and Los Angeles and Santa Barbara counties.

The CCC is exploring the possibility of participating in a video to be done by the Long Beach Fire Department, teaching residents about the proper methods of filling and using sandbags. The CCC is also exploring the possibility of holding floodfighting workshops with various cities, most likely in conjunction with the Department of Water Resources.

The CCC is working to develop and have contracts in place for emergency work with local governments.

CCC staff, including the Corps' Emergency Manager, have already become involved in planning meetings with various agencies, including the Department of Water Resources, the Office of Emergency Services and the Army Corps of Engineers.

Public Information Contact:

Susanne Levitsky
916/341-3145

Department of Fish and Game (DFG)

Actions to be Taken to Prepare for El Niño:

In light of the large number of notifications due to concerns for El Niño, The Department of Fish and Game (DFG) has continued to process streambed alteration agreements within statutory requirement.

DFG is providing training for staff for mutual aid response and is continuing to repair facilities and equipment from the 1997 floods.

Actions to be Taken to Prepare for El Niño:

DFG El Niño preparedness includes stockpiling sand and gravel, examining hatchery operations to ensure fish can be moved from hatcheries that may be impacted by high flows or water quality problems. The fish would wither be planted or relocated to other hatcheries. DFG is training staff for mutual aid and specialized equipment and is gearing up an internal incident command.

Public Information Contact:

Jack Edwards
916/653-7664

Department of Forestry and Fire Protection (CDF)

CDF will provide hand crews, for a variety of work such as levee construction and can provide full-service response as from a fire department, including search and rescue, medical treatment, extrication, and hazardous materials cleanup. CDF has peace officers and trained crisis communicators, incident command teams, and large scale logistical and planning support.

Department of Parks and Recreation

The Department of Parks and Recreation is developing a program to make its lifeguards available to assist in search and rescue.

The Department of Parks and Recreation is clearing and repairing drainage systems; purchasing supplies and emergency equipment; clearing and repairing drainage systems; purchasing supplies and emergency equipment; moving portable lifeguard towers off beaches; securing campground and beach furniture; evaluating hillsides and retaining walls for potential slides; constructing berms at important beach and other locations; and winterizing historic structures.

Department of Conservation

The Division of Mines and Geology responds to landslide emergencies, coordinating its response with the OES. In serious situations, the State Geologist contacts the Department director's office. Division geologists report to the OES on field reviews, local geologic conditions, specific landslide hazards, conclusions and recommendations, and provide maps. The Public Affairs Office is alerted to all significant events and contacts news media as warranted.

Public Information Contact:

Chris Chrystal
916/653-6560

YOUTH & ADULT
CORRECTIONAL AGENCY

Actions Completed for Emergency Preparedness:

The Department of Corrections (CDC) and the Department of Youth Authority (DYA) have established an Emergency Operations Plan, to allow prison staff to adequately manage the impact of a disaster. These actions include sandbagging, providing transportation services, medical services, housing of displaced citizens, providing food services, supplying sanitation and potable water, providing peace officer personnel, facilitating hazardous material containment, and providing fire services.

Actions Taken to Prepare for El Niño:

During a disaster, the Department of Corrections can provide 182 emergency fire crews and the Department of the Youth Authority can provide 24 emergency fire crews which would be available for community work such as construction, cleanup, levee repair and provision of rescue services. In addition, a large work force from these two departments would be available to fill sandbags to be placed on pallets and then trucked to the disaster sites.

Emergency transportation of flood victims or emergency crews could be provided using State vehicles from each prison and the fleet of buses utilized to transport inmates within the CDC system. Two ward transportation buses and ten vans would also be available from the Department of the Youth Authority.

Within the CDC, there are four certified Acute Care Hospitals, as well as a Skilled Nursing Facility, which could provide medical services to displaced members of the community. Each of the 11 Youth Authority institutions could provide emergency medical care as well.

Many prison warehouses and other buildings outside the secure perimeter of a prison could be converted for temporary shelter for displaced community members.

Food services can be provided to displaced citizens via utilization of existing prison services and stores or in the field via utilization of CDC Conservation Camp Field Kitchens and assigned crews.

Most prisons have their own dedicated water supply and waste water treatment facilities which could be utilized to supplement services within the community that had been jeopardized or rendered unusable by flood waters or related disasters.

Each prison and Youth Authority institution have sworn peace officer personnel that could be utilized by the County Mutual Aid Coordinator (County Sheriff) to provide additional law enforcement services within the community.

Each prison has a trained Hazardous Materials Coordinator and Hazardous Materials Team and each Youth Authority institution has a Hazardous Materials Team that could be utilized to bolster efforts within the community to contain hazardous materials.

Each prison has trained fire suppression staff who routinely provide services to the community in a mutual aid capacity to bolster fire suppression efforts.

Public Information Contacts:

Department of Corrections	Tip Kindel	Department of the Youth Authority	J.P. Tremblay
	Christine May		916/262-1479
	Kati Couraut		
	916/445-4950		

OES Headquarters

2800 Meadowview Road
Sacramento, California 95832
916/262-1800
916/262-1677 FAX

Regional OES Offices

Coastal Region

1300 Clay Street, Suite 408
Oakland, California 94612
510/286-0895
510/286-0853 FAX

Inland Region

2800 Meadowview Road
Sacramento, California 95832
916/262-1772
916/262-2869 FAX

Southern Region

11200 Lexington Drive, Building 283
Los Alamitos, California 90720-5002
310/795-2900
310/795-2877 FAX

Welcome to CERES

The California Environmental Resources Evaluation System

http://ceres.ca.gov/

The California Environmental Resources Evaluation System - CERES - is an information system developed by the **California Resources Agency** to facilitate access to a variety of electronic data describing California's rich and diverse environments. The goal of CERES is to improve environmental analysis and planning by integrating natural and cultural resource information from multiple contributors and by making it available and useful to a wide variety of users.

Environmental Information:
- By Organization
- By Geographic Area
- By Theme
- By Data Type

Other Services:
- Help Desk
- Program Information

Current Interest:
- El Niño Information

CERES Webs:
- Environmental Education
- Environmental Law
- Land Use Planning Information Network
- Watershed Information Technical System
- Wetlands Information System

CERES Web Search: [ ] Search Search Options

| California Homepage | Comment | Disclaimer |

CALIFORNIA
El Niño
INFORMATION

http://ceres.ca.gov/elnino/

Lead El Niño Agencies in California
California government agencies and their roles in preparing for an El Niño winter

Current Information
The latest information on this El Niño phenomenon

What is El Niño?
Scientific descriptions, explanations, and comparisons to past El Niño events

Forecasts and Predictions
How El Niño might affect our short-term climate

Impacts and Effects
El Niño's possible long-term impacts on California and other regions

Emergency Preparedness & Recovery
Maps and Imagery

Additional Resources:
CERES Flood Information
CERES Climate and Weather
CERES Weather and Road Conditions by County
Educational Resources
Yahoo Current Events: El Niño

| CERES Search | Comment | CERES Home |

http://wwwdwr.water.ca.gov/

DWR Organizations

- *Mission*
- *Overview*
- *Organization*

State Water Project

- *Overview*
- *Operations*
- *Environmental Compliance*

Local Assistance

- *Division of Planning and Local Assistance*

What's New At DWR ?

- *News Releases*
- *Director's Water Commission Reports*
- *Water Calendar*

California Water Information

- *Surface Water*
- *Groundwater*
- *River Forecast, Reservoir Info*
- *Publications*
- *California Water Plan*

HOME PAGES

- *Other DWR Pages*
- *Other Agencies*
- *CERES - California Environmental Resources Evaluation System*
- *CDEC - California Data Exchange Center*
- *CALFED Bay-Delta Program*

California Department of Water Resources

DWR El Niño Info

For more information about **California Department of Water Resources** water activities write or phone the DWR Office of Water Education

Our New Look

Please read this important Disclaimer

Webmaster: Larry Filby

Text-Only Users

O E S
CALIFORNIA
Governor's Office of
Emergency Services
http://www.oes.ca.gov/
Governor's Office of Emergency Services
Winter Weather and Flood Information
Emergency Public Information
Departments
Facts About OES
California Emergency Plan
Related Agencies
Ready
To Ride It Out?
California Earthquake Preparedness

APEX
TP_CF
9780160560446